NEUROSCIENCE RESEARCH PROGRESS

NEW DEVELOPMENTS IN SEROTONIN RESEARCH

NEUROSCIENCE RESEARCH PROGRESS

Additional books in this series can be found on Nova's website under the Series tab.

Additional e-books in this series can be found on Nova's website under the e-book tab.

NEUROSCIENCE RESEARCH PROGRESS

NEW DEVELOPMENTS IN SEROTONIN RESEARCH

MING D. LI, PHD

EDITOR

New York

Library of Congress Cataloging-in-Publication Data

ISBN: 978-1-63482-220-6
Library of Congress Control Number: 2015932294

Published by Nova Science Publishers, Inc. † New York

Contents

Preface

Serotonin (5-HT) is an ancient molecule that appeared very early in evolution and is produced by organisms in several phyla. First identified in 1937 as the secretory product of enterocromaffin cells, later, serum-constricting substance was discovered, and it was eventually found that these two substances were the same molecule, 5-hydroxytryptamine. Only in recent years has 5-HT been considered one of the most evolutionarily fascinating molecules, being implicated in many biological processes and attracting studies in many fields, from neurobiology to developmental biology and neuropsychiatry. 5-Hydroxytryptamine is a polyfunctional signaling molecule, acting in both developing and maturing animals as a neurotransmitter in both the central and peripheral nervous systems and affecting endocrine hormones and growth factors.

Serotonin's diverse effects are mediated by several types of receptors distributed throughout the body. To date, at least 14 serotonin receptor subtypes have been identified in mammals and are grouped into seven families. All of the receptors are G protein-coupled receptors. They mediate both excitatory and inhibitory neurotransmission. Through activation with serotonin, the receptors modulate the release of many neurotransmitters, including glutamate, GABA, dopamine, epinephrine, norepinephrine, and acetylcholine, as well as many hormones, including oxytocin, prolactin, vasopressin, cortisol, corticotropin, and substance P. The serotonin receptors have been demonstrated to influence many biological and neurological processes such as aggression, anxiety, appetite, cognition, learning, memory, mood, nausea, sleep, and thermoregulation. Thus, they are the target of a variety of pharmaceutical agents, including many antidepressants, antipsychotics, anorectics, antiemetics, gastroprokinetic agents, antimigraine agents, hallucinogens, and entactogens.

The primary objective of this book is to provide readers with an up-to-date knowledge of advances in serotonin research in different fields. To reach this goal, a group of scientists who have been engaged in serotonin and serotonin receptor research with different approaches and organisms was invited to contribute. The research topics covered include the following. The first chapter, by Yang and Li, provides novel evidence, primarily from human genetic studies, supporting the involvement of genetic variants in the serotonin transporter and receptor genes in the etiology of nicotine dependence and alcoholism through epistasis, although no or only a weak association of these variants with addictions was revealed. In the second chapter, Opdal provides an update on genes and genetic variants in the serotonergic network that have been implicated in sudden infant death syndrome. The genes discussed in this chapter include serotonin transporter, serotonergic receptors, as well as genes encoding other components in the serotonergic network. The third chapter, authored by Alesidze and Nesil, discusses how aggressive social stresses such as the absence of electricity, cold weather, food shortages, and social isolation influence the development of chronic stress, pathological aggression, and the formation of killer and self-killer organisms and how the distribution and concentration of serotonin in the brain change along with these physiological and pathological states. Two chapters then focus on recent developments in serotonin research in *Xenopus* and fish. In Chapter 4, based on new discoveries in *Xenopus*, Ori et al. show that serotonin signaling influences many developmental processes both in the central nervous system and in the peripheral organs, indicating that the selective modulation of the activities of the serotonin receptor could be an attractive target for future therapeutic approaches. On the other hand, in Chapter 5, Kuz'mina reviews new evidence on the involvement of the brain's serotonergic system in the regulation of food intake, digestion, and metabolism in certain fish. Finally, in the last two chapters, Mekhtieve et al. describe a novel serotonin-modulating anticonsolidation protein (SMAP), which was first identified in the rat brain cortex and purified from the whole brain, with a molecular weight of 186 kDa. Through a series of behavioral and molecular studies with different experimental animals, the authors showed how this protein is involved in memory formation and suppression of drug addiction and epileptic seizures in chapter 6 and in regulation of mutagenesis and adaptation in chapter 7.

Together, these chapters provide a broad view of recent advances in serotonin research in different species and various fields. It is our hope that the book can provide readers with a greater understanding of these new developments. I am most grateful to the distinguished group of researchers who have come together to produce this book. These experts, united in their mission to deliver a scholarly and comprehensive book, come from animal and human research fields. I am grateful for all that they taught me through their contributions and for the knowledge they will convey to all who read this book.

Virginia, USA & Zhejiang, China; November 2014 Ming D. Li, PhD

In: New Developments in Serotonin Research ISBN: 978-1-63482-220-6
Editor: Ming D. Li

Chapter 1

Significant Contribution of Variants in Serotonin Transporter and Receptor Genes to Drug Addiction Through Gene-by-Gene Interaction

Zhongli Yang and Ming D. Li
Department of Psychiatry and Neurobehavioral Sciences
University of Virginia, Charlottesville, VA, US

Abstract

Both nicotine and alcohol addiction are common chronic brain disorders that are extremely costly to individuals and society. Although genetic factors contribute significantly to these disorders, the susceptibility genes and variants underlying them remain largely unknown. Many years of genome-wide and candidate gene-based association studies have implicated a number of genes and biological pathways in the etiology of nicotine and alcohol addictions. In this chapter, we focus on current evidence, primarily from human genetic studies, supporting the involvement of variants in the serotonin transporter and receptor genes in the etiology of nicotine dependence and alcoholism based on both individual SNP- and haplotype-based association analysis, as well as gene-by-gene interaction studies. Current efforts aim not only to replicate these findings in independent samples, but also to identify which variant(s) contributes to the etiology of nicotine dependence and alcoholism and through what molecular mechanisms.

Keywords: African Americans, alcoholism, epistasis, European Americans, 5-HT, 5-HT3 receptors, *HTR3A*, *HTR3B*, serotonin, *SLCA4*, smoking dependence, SNPs

Genetic Epidemiology of Nicotine Dependence and Alcoholism

Drug addiction is a serious public health concern. According to the World Health Organization (WHO, 2008), there were an estimated 2 billion alcohol abusers, 1.3 billion tobacco users, and 230 million illicit drug users worldwide in 2004. There is considerable evidence from family, twin, and adoption studies for the operation of genetic factors in the vulnerability to addiction and that genetic factors contribute substantially to inter-individual vulnerabilities, with an estimated moderate-to-high heritability for both nicotine dependence and alcoholism (Goldman et al., 2005; Ho et al., 2010; Li and Burmeister, 2009; Li et al., 2003).

Estimates of the heritability of alcohol abuse and dependence range from 50%-70%. Additional studies have revealed a similarly high level of heritability across other alcohol-related behaviors, including heavy consumption and "problem" drinking (Gelernter and Kranzler, 2009; Goldman et al., 2005; Ho et al., 2010). For example, biological offspring of alcoholics are three to five times more likely to develop alcoholism than are the offspring of non-alcoholics. Most adoption studies show absence of a higher risk for alcoholism in adopted children of alcoholic foster parents, and the concordance rate for alcoholism in monzygotic twins is twice that in dizygotic twins. Similarly, many large twin studies have concluded that genetics contributes significantly to the risk of becoming a regular and dependent smoker (Heath et al., 1993; Heath et al., 1998; Heath and Martin, 1993; Kendler et al., 1999; Swan et al., 1990; True et al., 1997; True et al., 1999). Meta-analysis of a dozen twin studies showed that both genetics and environment play important roles in smoking-related behaviors, with an estimated heritability for nicotine dependence of 0.59 in male and 0.46 in female smokers, an average of 0.56 for the population as a whole (Li et al., 2003). In addition, nicotine dependence is influenced by environmental factors, as well as by gene–gene and gene–environment interactions (Ho et al., 2010; Lessov-Schlaggar et al., 2008; Li et al., 2003; Sullivan and Kendler, 1999; Swan et al., 2003).

Serotonin Transporter and Receptor Genes in Humans

Serotonin (5-hydroxytryptamine; 5-HT) is a neurotransmitter that mediates rapid excitatory responses through specific receptors (i.e., 5-HT3 receptors). The 5-HT3 receptors, unlike other serotonergic receptor classes, which are G protein-coupled (Barnes et al., 2009; Boess and Martin 1994; Cravchik and Goldman 2000), belong to the superfamily of nicotinic acetylcholine (nACh), subtype A of the γ-aminobutyric acid ($GABA_A$) and glycine receptors (Maricq et al., 1991).

The serotonin-gated ion channel conducts primarily Na^+ and K^+, resulting in rapid neuronal depolarization followed by rapid desensitization and the release of stored neurotransmitter, suggesting an important role for this receptor system in neuronal circuitry involved in drug abuse and addiction (Grant 1995).

Further, 5-HT3 receptors are co-localized with nACh receptors on nerve terminals in several brain pathways involved in reward processing, including dopaminergic terminals in the striatum (Nayak et al., 2000). Although there is no evidence that they interact physically, cross-regulation may take place at a downstream molecular level (Dougherty and Nichols 2009; Nayak et al., 2000; Yamauchi et al., 2011). Besides the potentially important role of 5-HT3 receptors in the development of nicotine dependence, they can be potentiated through acute exposure to alcohol at concentrations that produce intoxication (Narahashi et al., 2001; Sung et al., 2000).

Whereas 5-HT3 receptors assembled by 5-HT3_A subunits are uniformly located in various parts of the central and peripheral nervous systems, transcripts of the 5-HT3_A and 5-HT3_B subunits are co-expressed in the amygdala, caudate, and hippocampus, areas implicated in alcohol, nicotine, and other drug addictions, and form pharmacologically more potent heteropentameric receptors than the 5-HT3_A homomeric structures (Davies et al., 1999; Dubin et al., 1999; Enoch et al., 2011). The genes encoding the 5-HT3_A and 5-HT3_B receptor subunits (namely, *HTR3A* and *HTR3B*) lie in a 90-kb region on chromosome 11q23.1 (Miyake et al., 1995).

Serotonin transporters, one major class of monoamine transporters, which regulate the availability of 5-HT in the synaptic cleft through re-uptake, are encoded by the *SLC6A4* gene on chromosome 17q11.2 (Ramamoorthy et al., 1993). *SLC6A4* spans 37.8 kb and is composed of 14 exons (Lesch et al., 1994). The protein encoded by this gene, 5-HTT, is a transmembrane protein containing 630 amino acids (Heils et al., 1996). The expression of *SLC6A4* is regulated by at least three mechanisms: transcription regulatory elements in the promoter (Ramamoorthy et al., 1993), differential splicing (Bradley and Blakely, 1997), and the use of different 3′-polyadenylation sites (Battersby et al., 1999).

Of the variants investigated in *SLC6A4* for addictions and other psychiatric disorders, the long (L) and short (S) polymorphism at the 5-HTT-linked polymorphic region (5-HTTLPR) is the most extensively studied one. The 5-HTTLPR locates in the 5-HTT transcriptional control region and contains 16 tandem repeats of a 20- to 23-bp (G + C)-rich sequence between basepairs -1376 and -1027 from the transcriptional start site. Two common forms of this transcriptional control region have been found: a long 528-bp allele (L) with 16 repeats and a short 484-bp allele (S) with a deletion of 44 bp extending from basepairs -1255 to -1212. Although the 5-HTTLPR polymorphism of *SLC6A4* has been extensively studied, the results are inconclusive. For example, in a meta-analysis of 17 studies, Feinn et al., (2005) showed that the S allele was significantly associated with alcohol dependence in subjects with co-occurring serotonergic abnormalities, whereas several other studies reported an association of alcohol dependence with the L allele (Kweon et al., 2005, Hu et al., 2005). On the other hand, numerous studies reveal a differential association between chronic problem drinking and the density and function of serotonin transporters in alcoholic subjects carrying L and S variants of SLC6A4 (Little et al., 1998, Javors et al., 2005, Johnson et al., 2008).

Furthermore, several other polymorphisms that change the amino acid sequence, such as Thr4Ala, Gly56Ala, Glu215Lys, Lys605Asn, and Pro612Ser, of 5-HTT affect 5-HT uptake function in cell cultures (Prasad et al., 2003). Sequence variations in *SLC6A4* have been associated with several neuropsychiatric conditions, including major depressive disorders, anxiety-related personality traits, and antidepressant response (Dong, Wong, and Licinio 2009; López-León et al., 2007; McCauley et al., 2004).

Contribution of Variants in Serotonin Transporter and Receptor Genes to Nicotine Addiction

Nicotine, the primary psychoactive ingredient in cigarette smoke, exerts its pharmacologic and physiological effects by readily crossing the blood–brain barrier and binding with nicotinic acetylcholine receptors (nAChRs) located in various brain structures (Wonnacott, 1997). Activation of nAChRs on dopaminergic terminals induces dopamine release in the mesolimbic brain reward system (Kleijn et al., 2011; Wonnacott et al., 2000). In addition to dopamine, nicotine can induce the release of a variety of other neurotransmitters that regulate the complex brain circuitry modulating various addictive behaviors (Mihailescu et al., 1998; Ribeiro et al., 1993). Of these neurotransmitter systems, the serotonin system regulates self-administration of nicotine (Levin et al., 2008), nicotine-induced cognitive enhancement (Levin and Rezvani, 2007), memory (Levin et al., 2005), and smoking progression and cessation (Wang and Li, 2010).

To determine whether genetic variants in *SLC6A4*, *HTR3A,* and *HTR3B* contribute to the etiology of nicotine dependence, Yang et al. (2013) genotyped two SNPs (5'-HTTLPR and rs1042173) in *SLC6A4*, 8 SNPs in *HTR3A*, and 7 SNPs in *HTR3B* in a sample consisting of 1,366 African Americans (AAs) representing 402 nuclear families and 671 European Americans (EAs) representing 200 families and analyzed their association with nicotine dependence, which was assessed by smoking quantity (SQ), heaviness of smoking index (HSI), and Fagerström test for nicotine dependence. Individual SNP-based association analysis revealed marginal associations of rs10160548 in *HTR3A* with SQ and HSI ($P = 0.030$ and 0.042, respectively) in the AA sample, rs11606194 in *HTR3B* with SQ and FTND ($P = 0.039$ and 0.028, respectively) in the pooled AA and EA samples, and 5-HTTLPR in *SLC6A4* with FTND ($P = 0.03$) in EAs.

Following individual SNP-based association analysis, Yang and colleagues (2013) performed haplotype-based association analysis with the same genotyping data in the same samples, which revealed the following main findings. In AAs, there were two major haplotypes located in the 5′ region of *HTR3A* that were significantly associated with the three ND measures: (1) G-C-C-T-A-T, formed by SNPs rs1150226, rs1062613, rs33940208, rs1985242, rs2276302, and rs10160548 (LD block 3; Figure 1), with a frequency of 19.5%, was associated significantly with SQ ($Z = 2.596$; $P = 0.009$), HSI ($Z =$

3.027; P = 0.002), and FTND (Z = 2.824; P = 0.004) in a dominant model; and (2) G-A, formed by SNPs rs1150220 and rs1176713 (LD block 4; Figure 1), with a frequency of 66.6%, was significantly associated with SQ (Z = 3.041; P = 0.002,), HSI (Z = 3.011; P = 0.003), and FTND (Z = 2.863; P = 0.004). All these haplotype-based associations remained significant after Bonferroni correction for multiple testing for each LD block. We also detected a nominally significant association of haplotype A-T-C-A-G-G in the LD block 3 of *HTR3A* with a frequency of 24.1% with SQ under the dominant model (Z = 1.996; P = 0.046).

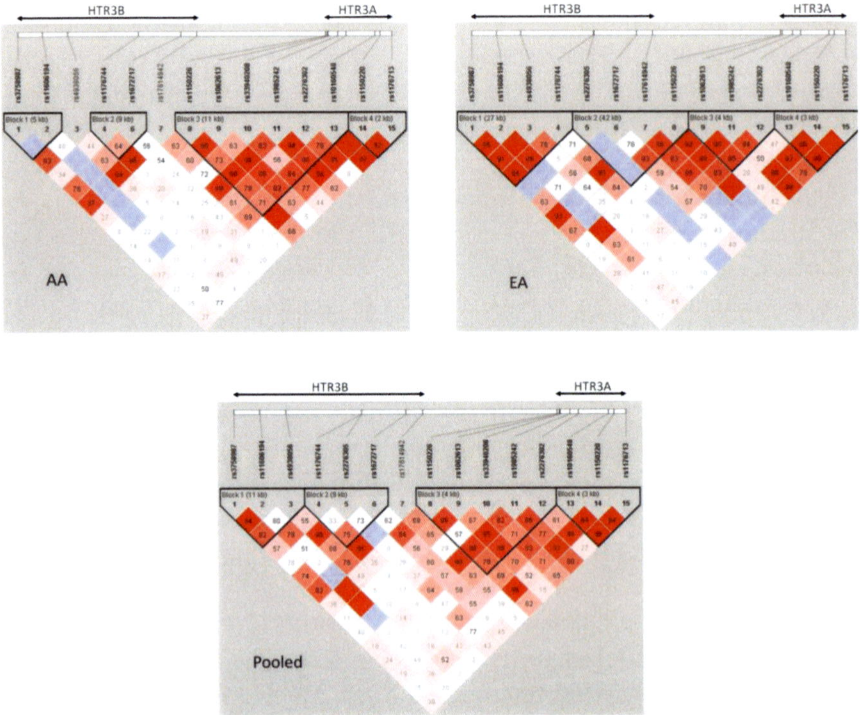

Figure 1. LD structure for *HTR3B* and *HTR3A* SNPs in the AA, EA, and pooled samples of Mid-South Tobacco Family study. Haploview was used to calculate all D' values and haplotype blocks were defined according to Gabrial et al. (2002). The number in each box represents the D' value for each SNP pair surrounding that box. Reprinted from Yang et al. (2013): Serotonin transporter and receptor genes significantly impact nicotine dependence through genetic interactions in both European American and African American smokers. *Durg and Alcohol Dependence* 129: 217-225, with permission from Elsevier.

For the EA sample, we found no haplotypes in *HTR3A* or *HTR3B* showing significant association with ND. In the pooled AA and EA sample, we found one haplotype, G-G-G, formed by SNPs rs10160548, rs1150220, and rs1176713 of *HTR3A*, with a frequency of 13.5%, significantly associated with SQ (Z = -2.377; P = 0.017), HSI (Z = -2.310; P = 0.021), and FTND (Z = -2.190; P = 0.029). However, none of them remained significant after Bonferroni correction.

Contribution of Variants Serotonin Transporter and Receptor Genes to Alcoholism

Alcoholism is a complex multifactorial disorder with a non-Mendelian pattern of inheritance. Findings from family, twin, and adoption studies have consistently demonstrated a significant role of genetics in the development of alcoholism (Dick and Bierut, 2006; Gelernter and Kranzler, 2009; Goldman et al., 2005). Numerous studies have demonstrated the significance of opioids (i.e., β-endorphin), dopamine (DA), serotonin (5-HT), γ-amino-butyric acid (GABA), and glutamate for the development and maintenance of alcohol dependence. Further, serotonin receptors contribute to alcohol use in animals, as alcohol increases basal concentrations of 5-HT affecting receptors (Kelai et al., 2008; Ward et al., 2009). Of the seven distinct families of 5 HT receptors, three contribute to alcohol dependence: 5-HT1A receptors might be associated with alcohol consumption and the development of tolerance, 5-HT2 receptors with reward, and 5-HT3 receptors with the development of reinforcement.

The 5-HT3 receptor is also involved in the expression of alcohol's rewarding effects. Behavioral pharmacological studies show that many of alcohol's rewarding effects are mediated by interactions between dopamine and 5-HT receptors in the midbrain and cortex. The 5-HT receptors are densely distributed in the terminals of mesocorticolimbic dopamine-containing neurons, where they regulate dopamine release in these brain regions. These dopamine pathways, particularly those in the nucleus accumbens, are critically involved in mediating the rewarding effects of alcohol.

Genetically, to determine whether variants in serotonin transporters and receptors contribute to the development of alcohol addiction, Seneviratne et al. (2013) analyzed 22 SNPs from *SLC6A4*, *HTR3A*, and *HTR3B* for their association with alcohol dependence (AD) in a Caucasian case-control sample.

Association analysis of individual SNPs with AD revealed that four SNPs in *HTR3B* (i.e., rs3758987, rs4938056, rs3782025, and rs1672717) and the 5′-HTTLPR in *SLC6A4* are marginally associated with AD (Seneviratne et al., 2013). Of these, only the association of AD with rs3782025 in *HTR3B* remained significant after correcting for multiple comparisons with the SNPSpD program (Nyholt, 2004). Among the three relatively low-frequency SNPs examined, minor alleles of both rs33940208 in *HTR3A* and rs2276305 in *HTR3B* were significantly protective against AD with odds ratios (ORs) of 0.212 (95% confidence interval [CI] 0.073, 0.616; P = 0.004) and 0.261 (95% CI 0.088, 0.777; P = 0.016), respectively.

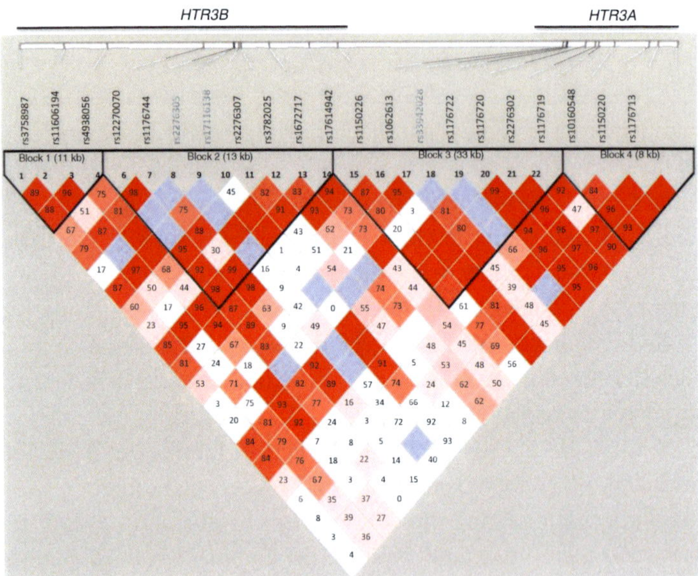

Figure 2. LD structure for HTR3B and HTR3A SNPs and haplotype blocks for alcohol study. Less-frequent variant SNPs are indicated in ligher color font.

Next, Seneviratne and colleagues (2013) performed linkage disequilibrium (LD) analysis for these SNPs in the same case-control Caucasian sample, which revealed two LD blocks within *HTR3A* and two LD blocks within *HTR3B* (Figure 2). In *HTR3A*, neither the haplotypes formed by the six SNPs rs17614942, rs1150226, rs1062613, rs1176722, rs1176720, and rs2276302 within Haploview-defined "LD Block 3", nor the haplotypes formed by the four SNPs rs1176719, rs10160548, rs1150220, and rs1176713 within "LD Block 4" were significantly associated with AD. Within LD Block 3, the low-frequency SNP rs33940208 and the common SNP rs1062613, located 71 bp upstream of rs33940208, formed a two-SNP haplotype C-T with a nominally significant protective effect against AD (haplotype score -2.2259; P value for individual haplotype association analysis 0.0260, and global P = 0.0347; OR = 0.3706; 95% CI 0.1283, 1.0705).

In *HTR3B,* a 5-SNP haplotype G-A-A-A-A formed by SNPs rs12270070, rs1176744, rs2276307, rs3782025, and rs1672717 in Haploview-defined “LD Block 2” was found to be significantly associated with AD (haplotype score -2.693; P value for individual haplotype association analysis 0.007, and global P = 0.0004; OR = 0.785; 95% CI 0.598, 1.031). The LD Block 2 is located in and around four exons that are common to all mRNA isoforms of *HTR3B*.

Contribution of Variants in Serotonin Transporter and Receptor Genes to Addictions Through Gene–Gene Interaction

Like any other complex trait and as described in previous sections of this chapter, addiction is controlled by multiple genetic factors, with each having a relatively modest effect, and by environmental factors, as well as by both gene–gene (epistatic) and gene–environment interactions (Flint and Munafo, 2008; Ho et al., 2010; van der Zwaluw and Engels, 2009). As documented in the literature (Gelernter and Kranzler, 2009; Ho and Tyndale, 2007; Lessov-Schlaggar et al., 2008; Li, 2008; Li and Burmeister, 2009), significant efforts have been made in searching for vulnerability genes to addictions. However, these approaches are effective only for genes with moderate to major effects. The ability to identify susceptibility genes for addictions has been improving but remains considerably limited by the presence of various factors such as epistatic interaction, small to modest genetic effects, small sample size, and heterogeneities. Among these factors, detecting gene–gene and gene–environment interactions appears to be more challenging (Flint and Munafo, 2008; Ho et al., 2010; Lou et al., 2008; Lou et al., 2007; van der Zwaluw and Engels, 2009).

In the search for determinants of gene–gene and gene–environment interactions, significant efforts have been expended. Several combinatorial approaches, such as the multifactor dimensionality reduction (MDR) (Ritchie et al., 2001), the combinatorial partitioning method (CPM) (Nelson et al.,

2001), and the restricted partition method (RPM) (Culverhouse et al., 2004), represent promising tools for detecting gene–gene and gene–environment interactions. Since the original report, MDR has been one of the most widely applied approaches to detect interactions underlying a spectrum of complex disorders. However, these established methods have critical limitations that restrict their practical use. For example, MDR, CPM, and RPM do not allow adjustments for covariates; MDR is applicable only to dichotomous phenotypes; and CPM and RPM cannot handle categorical phenotypes. To overcome the limitations of these combinatorial approaches and to meet research needs in determining gene–gene and gene–environment interactions for complex phenotypes, a generalized MDR (GMDR) as well as a pedigree-based GMDR (PGMDR) have been developed for case-control (Lou et al., 2007) and family-based (Lou et al., 2008) study designs, respectively, that permit adjustments for discrete and quantitative covariates and are applicable to both dichotomous and continuous phenotypes.

Epistatic effects of serotonin transporter and receptors on ND: Considering the biological and pharmacological functions of the three genes in regulating serotonin signaling, Yang et al. (2013) performed comprehensive interactive analysis among the 17 polymorphisms in *HTR3A*, *HTR3B*, and *SLC6A4* for their epistatic effect on the three ND measures in the same AA, EA, and pooled samples used for both the individual SNP- and haplotype-based association analysis described above. As shown in Table 1, the detected best interaction model for each sample shows a significant genetic interaction for all three ND measures, with an empirical $P < 0.01$, cross-validation consistency (CVC) of at least 7 of 10, and test accuracies (TA) > 50% based on 10^6 permutation tests except for the model on FTND in the AA sample, where the empirical P value is 0.057. Of the three samples, the epistatic effect of the detected best interaction model for the three ND measures in the pooled sample appeared to be the strongest, with an empirical P value of 0.00025 to 0.00085. In the pooled sample, an interaction model consisting of five polymorphisms—rs1062613 and rs10160548 in *HTR3A*, rs1176744 in *HTR3B*, and 5-HTTLPR and rs1042173 in *SLC6A4*—showed significant epistatic effects on all the three ND measures. Interestingly, the minor allele frequencies of these five polymorphisms are quite high, with the lowest frequency being 0.211 for rs1042173 in *SLC6A4* in the AA sample and 0.244 for rs1062613 in *HTR3A* in the EA sample. Of them, three polymorphisms have been demonstrated to alter the expression of the RNA or protein encoded by the respective genes (Niesler et al., 2001). For example, the rs1062613 is a translation regulatory variant located in an open reading frame upstream of the translation initiation site of *HTR3A* mRNA (Niesler et al., 2001).

Table 1. Detected best SNP combination of *SLC6A4*, *HTR3A*, and *HTR3B* associated with ND measures in EA, AA, and pooled sample on basis of test accuracy and empirical P value from 10^6 permutations

Sample	SNP Combination	ND Measure	Test Accuracy	Cross-Validation Consistency (CVC)	Permutated P value
EA	*HTR3A*: rs1062613, rs1150220; *HTR3B*: rs1176744; *SLC6A4*: 5-HTTLPR, rs1042173	SQ	0.5678	7	0.003
		HSI	0.5699	9	0.002
		FTND	0.5703	10	0.002
AA	*HTR3A*: rs10160548; *SLC6A4*: 5-HTTLPR, rs1042173	SQ	0.5500	10	0.005
		HSI	0.5458	10	0.009
		FTND	0.5317	8	0.057
Pooled	*HTR3A*: rs1062613, rs10160548; *HTR3B*: rs1176744; *SLC6A4*: 5-HTTLPR, rs1042173	SQ	0.5516	8	0.00051
		HSI	0.5547	8	0.00025
		FTND	0.5479	10	0.00085

The two polymorphisms in *SLC6A4* were shown to alter 5-HTT expression through transcription regulation for 5-HTTLPR and degradation of mRNA transcripts for rs1042173 (Heils et al., 1997; Heils et al., 1996; Seneviratne et al., 2009; Vallender et al., 2008). Of the remaining SNPs, rs10160548 is located in intron 6 near an intron–exon boundary. It is thus reasonable to speculate that it alters the expression of functional *HTR3A* transcripts through alternative splicing. The rs1176744 in *HTR3B* does not alter expression, but it substantially changes serotonergic signaling through altered gating kinetics of the 5-HT_{3AB} receptor complex (Krzywkowski et al., 2008). By analyzing the AA and EA samples independently, Yang et al. (2013) revealed slightly different interaction models for each ethnic sample. In the AA sample, there was a significant interactive effect of polymorphisms rs10160548 in *HTR3A* and 5-HTTLPR and rs1042173 in *SLC6A4* on all three ND measures. Although the two *SLC6A4* polymorphisms were also included in the best interaction model detected in the EA sample, the model contained three additional loci: rs1062613 and rs1150220 in *HTR3A* and rs1176744 in *HTR3B*. In previously reported studies, SNP rs1062613 in *HTR3A* was associated with several psychiatric disorders in individuals of European descent (Gatt et al., 2010; Walstab et al., 2010). Yet whether rs1062613 has ethnicity-specific *cis*-acting effects on the differential translation levels of *HTR3A* in AAs and EAs remains to be characterized. However, inclusion of rs1062613 in the best interaction model in the pooled samples, with even stronger interaction effects than were seen in EAs only, argues against this possibility. The other *HTR3A* SNP, detected only in the EA sample, was rs1150220, which is moderately correlated with rs10160548 in both EAs and AAs ($r^2 = 0.42$ in AAs and 0.51 in EAs) in an LD block located at the 3'-end of the *HTR3A* gene. The second main difference between AA and EA samples was the absence of *HTR3B* rs1176744 in the best model for AAs. Although SNPs rs1176744 and 5-HTTLPR in *SLC6A4* have been found to be significantly associated with alcohol dependence in AAs (Enoch et al., 2011), no significant association of these two polymorphisms with ND was revealed in their AA, EA, or pooled samples, except for 5-HTTLPR, which showed a marginal association with FTND in the EA sample. However, these genetic interaction analyses demonstrated that the two polymorphisms in *SLC6A4* play an important role in ND through interactions with other SNPs in *HTR3A* and *HTR3B* in the AA, EA, and pooled samples.

Epistatic effects of variants in serotonin transporter and receptors to AD: As shown in Table 2, Seneviratne et al. (2013) detected two four-variant interaction models with significant genetic interactive effects on AD, with an empirical $P < 10^{-6}$, cross-validation consistency (CVC) of 10, and TAs > 50% based on 10^6 permutation tests. Because three of the four variants included in the two interaction models were identical, the two four-variant models were combined, which formed a five-variant model. The five-variant interaction model had a significant genetic interactive effect on AD, albeit with a smaller P value than the two four-variant models, possibly because of the smaller sample used in the five-variant interaction model. Subsequent to detecting the interaction models, Seneviratne et al. (2013) analyzed the "genotype combinations" within the two four-variant and five-variant interaction models, as shown in Table 2. As shown clearly in Table 3, persons with all genotype combinations listed in the three detected best genotype combinations had a greater risk of AD than did those individuals carrying other genotype combinations, with ORs ranging from 2.37 (P = 0.0002) for the epistatic model 3 to 3.11 (P = 0.0001) for the epistatic model 2. Thus, our results provide evidence for strong genetic interactions within the serotonergic system underlying AD.

Table 2. Most significant SNP combinations of *SLC6A4*, *HTR3A*, and *HTR3B* detected by GMDR program for gene × gene interactions in Caucasian alcoholic sample

Epistatic Model	Combination	Test Accuracy	Cross-Validation Consistency (CVC)	P value*
1	*SLC6A4*: HTTLPR, rs1042173 *HTR3B*: rs1176744, rs3782025	0.8503	10	$<1\times10^{-6}$
2	*SLC6A4*: HTTLPR *HTR3A*: rs10160548 *HTR3B*: rs1176744, rs3782025	0.7402	10	$<1\times10^{-6}$
3	*SLC6A4*: HTTLPR, rs1042173 *HTR3A*: rs10160548 *HTR3B*: rs1176744, rs3782025	0.5836	9	0.0042

* Empirical P value calculated from 10^6 permutations.

Table 3. Best constituent genotype combination of *SLC6A4*, *HTR3A*, and *HTR3B* detected by GMDR program for gene × gene interactions in Caucasian alcoholic sample

Epistatic model	Combination	% in AD Cases (N=500)	% in Controls (N=280)	OR (95% CI)	P Value*
1	5HTTLPR: LL/LS; rs1042173: TT/TG; rs1176744: AC; rs3782025: AG	13.02	2.96	2.712 (1.578, 4.661)	0.0003
2	5HTTLPR: LL/LS rs10160548: GT/TT rs1176744: AC; rs3782025: AG	13.61	3.11	2.802 (1.648, 4.765)	0.0001
3	5HTTLPR: LL/LS; rs1042173: TT/TG rs10160548: GT/TT rs1176744: AC; rs3782025:AG	11.83	2.37	3.095 (1.715, 5.583)	0.0002

*Chi-square P values for comparisons between those carrying the genotype combination vs. non-carriers in AD cases and controls.

Comparison of Main Verse Interactive Effect of Variants in Serotonin Transporter and Receptors on ND

The most important finding from the study reported by Yang et al. (2013) is that none of the polymorphisms included in the epistatic models shown in Table 2 was significant at the individual locus level. Significant epistatic effects of variants without major genetic effect have become an increasingly identified phenomenon in studies of complex disorders (Li et al., 2008; Steen, 2012; Zuk et al., 2012). For example, several studies have shown marginal or no association of the polymorphism 5-HTTLPR in *SLC6A4* with ND (Gerra et al., 2005; Trummer et al., 2006; Yang et al., 2013). Yet, as clearly demonstrated by the gene-by-gene interaction analysis, the effect of 5-HTTLPR on ND is highly significant when its epistatic effect is taken into consideration. Another unique strength of these findings is that the interaction models detected in AAs, EAs, and the pooled samples were highly significant across multiple ND measures, providing further support for their role in smoking-related behaviors.

The biological basis for the genetic interaction effect we detected can be explained by the actions of nicotine on serotonergic signaling. Nicotine competes with its natural ligand serotonin for 5-HT$_3$ receptors (Breitinger et al., 2001). Depending on whether the 5-HT$_3$ receptors are located pre- or post-synaptically, nicotine binding can result in either the release of various neurotransmitters or changes in the propagation of fast-acting serotonergic signals along the post-synaptic neuron. The availability of synaptic serotonin for binding to the 5-HT$_3$ receptors is modulated by the presynaptic 5-HTTs. Prior studies have found mixed effects of chronic nicotine exposure on the density of 5-HTTs, thus regulating the amount of synaptic serotonin available for action on the 5-HT$_3$ receptors. For example, Semba and Wakuta (2008) reported a reduction in the density of 5-HTTs in the rat brain, while two other studies reported an elevation in 5-HTTs (Awtry and Werling, 2003; Slotkin and Seidler, 2010). On the other hand, Staley et al., (2001) reported an elevation of 5-HTTs in the human brain, whereas in human platelets, they were reported to be reduced (Patkar et al., 2003). Serotonin plays a crucial role in mediating cognitive behavioral functions, stress response, mood, appetite, and motor functions (Jasinska et al., 2012). Thus, the interactions among the three genes may represent the interacting biological effects of nicotine on fast-acting serotonergic signaling in ND.

Comparison of Main versus Interactive Effect of Variants in Serotonin Transporter and Receptors on AD

As shown in Table 3, the five interacting risk genotypes within *SLC6A4* (5′-HTTLPR:LL/LS; rs1042173:TT/TG), *HTR3A* (rs10160548:GT/TT), and *HTR3B* (rs1176744:AC; rs3782025:AG) are variants associated with alcoholism or related intermediate phenotypes. From Table 3, it is clear that these interacting genotypes are linked to about a three-fold greater risk of AD than their constituent individual genotypes. Importantly, allele-based functional differences for three of these five variants (i.e., *SLC6A4:* 5′-HTTLPR:LL/LS and rs1042173:TT/TG; *HTR3B*: rs1176744:AC) were demonstrated to be functional in earlier molecular studies. More specifically, alcoholics carrying the 5′-HTTLPR:L allele have lower serotonin transporter binding and reuptake activity than homozygous 5′-HTTLPR:S carriers (Johnson et al., 2008). Similarly, the rs1042173:T allele was associated with lower expression of serotonin transporters (Lim et al., 2006; Seneviratne et al., 2009). Thus, the 5′-HTTLPR:L and rs1042173:T alleles, in combination, may lead to a synaptic hyperserotonergic state. The third known functional variant *included in the risk genotype combinations was the* non-synonymous SNP *rs1176744 (A/C) in HTR3B, which results in a* tyrosine to serine change at the 129th amino acid residue in the extracellular N-terminal domain of 5-HT_{3B} (Walstab et al., 2008) close to the 5-HT_{3AB} interface (Krzywkowski et al., 2008). This amino acid substitution significantly increases 5-HT_{3AB} ion channel open time, augmenting serotonergic signaling (Krzywkowski et al., 2008). In another study, Enoch et al., (2011) reported a higher risk of AD associated with high activity of the serine allele in male African-American alcoholics. Further, several other studies reported associations of the *rs1176744* serine allele with other psychiatric disorders that often are co-morbid with AD in Caucasians (Frank et al., 2004; Hammer et al., 2012; Hammer et al., 2009; Yamada et al., 2006). It should be noted that, unlike the study by Enoch et al., in the study reported by Seneviratne et al., (2013), it was the heterozygous genotype of rs1176744 that contributed to all three risk genotype combinations and also showed a trend to risk at the single-locus level (OR 1.325; 95% CI 0.955, 1.838; P = 0.092). This suggests a molecular heterosis effect that could be race-specific. From a structural point of view, these findings can be interpreted as co-expression of two 5-HT_{3B} subunits in the same individual: one with the amino acid serine at position 129 and the other with tyrosine ("heterozygous" 5-HT_{3B} subunits) leading to a higher risk of AD. In such a situation, it is hard to decide which variant is more prevalent or the change of direction of serotonergic singling associated with the genotype. However, by using cultured tsA-201 cells (a transformed human kidney 293 cell line stably expressing an SV40 temperature-sensitive T antigen) without any treatment, Krzywkowski et al., (2008) demonstrated that expression of "heterozygous" 5-HT_{3B} subunits in the pentameric 5-HT_{3AB} receptor complex would mimic the heterozygous carrier state. Expression of "heterozygous" 5-HT_{3B} subunits resulted in an increase in the time the channel was open compared with 5-HT_{3AB} complexes expressing two homozygous wild-type 5-HT_{3B} subunits with the tyrosine residue (Hammer et al., 2009; Krzywkowski et al., 2008). Longer open time of the 5-HT_{3AB} channels can enhance the responsiveness of neurons to serotonin. This enhanced responsiveness associated with rs1176744:AG may be heightened by the greater availability of synaptic serotonin that is associated with the *SLC6A4:* 5′-HTTLPR:LL/LS and rs1042173:TT/TG genotypes.

Another important finding from the study reported by Seneviratne et al. (2013) was the protective effects against AD shown by the two relatively low-frequency synonymous variants rs33940208 in *HTR3A* exon 1 and rs2276305 in *HTR3B* exon 5. One caveat to these findings is that their significance levels did not reach the threshold for effectiveness in multiple testing. Nevertheless, these associations had medium to large effects in the primary analysis (Cohen's d for allelic associations of rs2276305:A and rs33940208:T with AD 0.74 and 0.86, respectively). The low-frequency variants with a protective effect on AD have not been reported previously in European populations. As both rs33940208 and rs2276305 are located within 100 basepairs of two well-characterized functional common variants (see Figure 2), they may modulate the functionality of the common variants. The rs33940208:T allele was seen only in carriers of rs1062613:C. The C allele of rs1062613 has been associated with lower *HTR3A* expression (Kapeller et al., 2008) and the CC genotype with high anxiety and greater amygdala responsiveness to emotional stimuli in humans (Iidaka et al., 2005; Kilpatrick et al., 2011). As high anxiety is a risk factor for AD, one would expect the protective allele to be linked to the non-risk T allele. On the other hand, it is tempting to speculate that the protective effects of the rs33940208:T allele might offset reductions in transcription associated with the rs1062613:C allele, constituting a subpopulation with greater expression than in the carriers of the rs1062613:C–rs33940208:G haplotype. Similarly, the other protective allele detected in the study reported by Seneviratne et al., (2013), rs2276305:A, was seen only in carriers of the major allele of the common functional SNP rs1176744:A, which previously has been reported as the non-risk allele (Enoch et al., 2011). Taken together, these findings support the view that functionality attributed to common variants is in fact a consequence of "synthetic associations" arising from the combined effects of multiple rare variants in their vicinity (Goldstein, 2009). In fact, several examples have been reported of similar rare-common SNP modulating effects on other common disorders with high heritability rates such as AD (Raychaudhuri, 2011). Thus, deep sequencing of *HTR3A* exon 1 and *HTR3B* exon 5 regions might help detect multi-variant modulating effects on the common functional SNPs rs1062613 and rs1176744.

Conclusion

In summary, by examining the association of variants in *HTR3A*, *HTR3B*, and *SLC6A4* with ND and AD at both the individual SNP and haplotype levels, only marginal association of variants in the three genes with the two addictive phenotypes was revealed in one of the studied samples. However, when these variants were examined interactively through gene–gene interaction approach, a combination of functional polymorphisms in the three genes with significant interaction effects on ND and alcoholism was identified. This strongly indicates these genetic variants play a significant role in both ND and AD through a genetic epistatic effect. This suggests it is important to investigate genetic epistatic effect when one searches susceptibility loci for a complex trait such as ND and AD, as described in this chapter. Importantly, SNPs rs10160548 in *HTR3A*, rs1176744 in *HTR3B*, and 5-HTTLPR and rs1042173 in *SLC6A4* were found to be significant, influencing both ND and AD through epistasis.

In spite of this progress in molecular genetic studies of addictions, we still have a long way to go, and there are many challenges that remain to be surmounted (Ho et al., 2010; Li, 2010; van der Zwaluw and Engels, 2009). These challenges include: (1) further identification and replication of known and unknown genes in the serotonin receptor and other signaling pathways and functional variants (including rare variants) for various addictive disorders through high-throughput approaches such as association study and deep sequencing analysis; (2) study of copy number variations (CNVs) and their impact on gene expression in serotonin signaling pathway and other addiction-related signaling pathways; (3) better understanding of the mechanisms underlying addictions at the molecular and cellular levels using both in vitro and in vivo approaches; and (4) determining appropriate ways of defining environmental factors such that we can assess how gene-environment interaction affects addictions. An improvement of our understanding of the genetic and environmental factors underlying drug addiction has considerable potential to reduce morbidity and death greatly by providing the most suitable methods for prevention and novel medications for treating different addictive disorders.

References

Awtry, T.L., Werling, L.L., 2003. Acute and chronic effects of nicotine on serotonin uptake in prefrontal cortex and hippocampus of rats. *Synapse* 50, 206-211.

Breitinger, H.G., Geetha, N., Hess, G.P., 2001. Inhibition of the serotonin 5-HT3 receptor by nicotine, cocaine, and fluoxetine investigated by rapid chemical kinetic techniques. *Biochemistry* 40, 8419-8429.

Culverhouse, R., Klein, T., Shannon, W., 2004. Detecting epistatic interactions contributing to quantitative traits. *Genet Epidemiol.* 27, 141-152.

Dick, D.M., Bierut, L.J., 2006. The genetics of alcohol dependence. *Curr. Psychiatry Rep.* 8, 151-157.

Enoch, M.A., Gorodetsky, E., Hodgkinson, C., Roy, A., Goldman, D., 2011. Functional genetic variants that increase synaptic serotonin and 5-HT3 receptor sensitivity predict alcohol and drug dependence. *Mol. Psychiatry* 16, 1139-1146.

Flint, J., Munafo, M.R., 2008. Forum: Interactions between gene and environment. *Curr. Opin. Psychiatry* 21, 315-317.

Frank, B., Niesler, B., Nothen, M.M., Neidt, H., Propping, P., Bondy, B., Rietschel, M., Maier, W., Albus, M., Rappold, G., 2004. Investigation of the human serotonin receptor gene HTR3B in bipolar affective and schizophrenic patients. *Am. J. Med. Genet. B Neuropsychiatr Genet.* 131B, 1-5.

Gabriel, S.B., Schaffner, S.F., Nguyen, H., Moore, J.M., Roy, J., Blumenstiel, B., Higgins, J., DeFelice, M., Lochner, A., Faggart, M., Liu-Cordero, S.N., Rotimi, C., Adeyemo, A., Cooper, R., Ward, R., Lander, E.S., Daly, M.J., Altshuler, D., 2002. The structure of haplotype blocks in the human genome. *Science* 296, 2225-2229.

Gatt, J.M., Williams, L.M., Schofield, P.R., Dobson-Stone, C., Paul, R.H., Grieve, S.M., Clark, C.R., Gordon, E., Nemeroff, C.B., 2010. Impact of the HTR3A gene with early life trauma on emotional brain networks and depressed mood. *Depress Anxiety* 27, 752-759.

Gelernter, J., Kranzler, H.R., 2009. Genetics of alcohol dependence. *Hum. Genet* 126, 91-99.

Gerra, G., Garofano, L., Zaimovic, A., Moi, G., Branchi, B., Bussandri, M., Brambilla, F., Donnini, C., 2005. Association of the serotonin transporter promoter polymorphism with smoking behavior among adolescents. *Am. J. Med. Genet. B. Neuropsychiatr. Genet.* 135, 73-78.

Goldman, D., Oroszi, G., Ducci, F., 2005. The genetics of addictions: uncovering the genes. *Nat Rev Genet* 6, 521-532.

Goldstein, D.B., 2009. Common genetic variation and human traits. *N. Engl. J. Med.* 360, 1696-1698.

Hammer, C., Cichon, S., Muhleisen, T.W., Haenisch, B., Degenhardt, F., Mattheisen, M., Breuer, R., Witt, S.H., Strohmaier, J., Oruc, L., Rivas, F., Babadjanova, G., Grigoroiu-Serbanescu, M., Hauser, J., Roth, R., Rappold, G., Rietschel, M., Nothen, M.M., Niesler, B., 2012. Replication of functional serotonin receptor type 3A and B variants in bipolar affective disorder: a European multicenter study. *Transl. Psychiatry* 2, e103.

Hammer, C., Kapeller, J., Endele, M., Fischer, C., Hebebrand, J., Hinney, A., Friedel, S., Gratacos, M., Estivill, X., Fichter, M., Fernandez-Aranda, F., Ehrlich, S., Rappold, G., Niesler, B., 2009. Functional variants of the serotonin receptor type 3A and B gene are associated with eating disorders. *Pharmacogenet Genomics* 19, 790-799.

Heath, A.C., Cates, R., Martin, N.G., Meyer, J., Hewitt, J.K., Neale, M.C., Eaves, L.J., 1993. Genetic contribution to risk of smoking initiation: comparisons across birth cohorts and across cultures. *J. Subst. Abuse* 5, 221-246.

Heath, A.C., Madden, P.A., Martin, N.G., 1998. Statistical methods in genetic research on smoking. *Stat. Methods Med. Res.* 7, 165-186.

Heath, A.C., Martin, N.G., 1993. Genetic models for the natural history of smoking: evidence for a genetic influence on smoking persistence. *Addict. Behav.* 18, 19-34.

Heils, A., Mossner, R., Lesch, K.P., 1997. The human serotonin transporter gene polymorphism--basic research and clinical implications. *J. Neural Transm* 104, 1005-1014.

Heils, A., Teufel, A., Petri, S., Stober, G., Riederer, P., Bengel, D., Lesch, K.P., 1996. Allelic variation of human serotonin transporter gene expression. *J. Neurochem.* 66, 2621-2624.

Ho, M.K., Goldman, D., Heinz, A., Kaprio, J., Kreek, M.J., Li, M.D., Munafo, M.R., Tyndale, R.F., 2010. Breaking barriers in the genomics and pharmacogenetics of drug addiction. *Clin. Pharmacol. Ther.* 88, 779-791.

Ho, M.K., Tyndale, R.F., 2007. Overview of the pharmacogenomics of cigarette smoking. *Pharmacogenomics* J 7, 81-98.

Iidaka, T., Ozaki, N., Matsumoto, A., Nogawa, J., Kinoshita, Y., Suzuki, T., Iwata, N., Yamamoto, Y., Okada, T., Sadato, N., 2005. A variant C178T in the regulatory region of the serotonin receptor gene HTR3A modulates neural activation in the human amygdala. *J. Neurosci.* 25, 6460-6466.

Jasinska, A.J., Chua, H.F., Ho, S.S., Polk, T.A., Rozek, L.S., Strecher, V.J., 2012. Amygdala response to smoking-cessation messages mediates the effects of serotonin transporter gene variation on quitting. *Neuroimage* 60, 766-773.

Johnson, B.A., Javors, M.A., Roache, J.D., Seneviratne, C., Bergeson, S.E., Ait-Daoud, N., Dawes, M.A., Ma, J.Z., 2008. Can serotonin transporter genotype predict serotonergic function, chronicity, and severity of drinking? *Prog. Neuropsychopharmacol. Biol. Psychiatry* 32, 209-216.

Kapeller, J., Houghton, L.A., Monnikes, H., Walstab, J., Moller, D., Bonisch, H., Burwinkel, B., Autschbach, F., Funke, B., Lasitschka, F., Gassler, N., Fischer, C., Whorwell, P.J., Atkinson, W., Fell, C., Buchner, K.J., Schmidtmann, M., van der Voort, I., Wisser, A.S., Berg, T., Rappold, G., Niesler, B., 2008. First evidence for an association of a functional variant in the microRNA-510 target site of the serotonin receptor-type 3E gene with diarrhea predominant irritable bowel syndrome. *Hum. Mol. Genet.* 17, 2967-2977.

Kelai, S., Renoir, T., Chouchana, L., Saurini, F., Hanoun, N., Hamon, M., Lanfumey, L., 2008. Chronic voluntary ethanol intake hypersensitizes 5-HT(1A) autoreceptors in C57BL/6J mice. *J. Neurochem.* 107, 1660-1670.

Kendler, K.S., Neale, M.C., Sullivan, P., Corey, L.A., Gardner, C.O., Prescott, C.A., 1999. A population-based twin study in women of smoking initiation and nicotine dependence. *Psychol. Med.* 29, 299-308.

Kilpatrick, L.A., Labus, J.S., Coveleskie, K., Hammer, C., Rappold, G., Tillisch, K., Bueller, J.A., Suyenobu, B., Jarcho, J.M., McRoberts, J.A., Niesler, B., Mayer, E.A., 2011. The HTR3A polymorphism c. -42C>T is associated with amygdala responsiveness in patients with irritable bowel syndrome. *Gastroenterology* 140, 1943-1951.

Kleijn, J., Folgering, J.H., van der Hart, M.C., Rollema, H., Cremers, T.I., Westerink, B.H., 2011. Direct effect of nicotine on mesolimbic dopamine release in rat nucleus accumbens shell. *Neurosci. Lett.* 493, 55-58.

Krzywkowski, K., Davies, P.A., Feinberg-Zadek, P.L., Brauner-Osborne, H., Jensen, A.A., 2008. High-frequency HTR3B variant associated with major depression dramatically augments the signaling of the human 5-HT3AB receptor. *Proc. Natl. Acad. Sci. USA* 105, 722-727.

Lessov-Schlaggar, C.N., Pergadia, M.L., Khroyan, T.V., Swan, G.E., 2008. Genetics of nicotine dependence and pharmacotherapy. *Biochem. Pharmacol.* 75, 178-195.

Levin, E., Icenogle, L., Farzad, A., 2005. Ketanserin attenuates nicotine-induced working memory improvement in rats. *Pharmacol Biochem. Behav.* 82, 289-292.

Levin, E.D., Rezvani, A.H., 2007. Nicotinic interactions with antipsychotic drugs, models of schizophrenia and impacts on cognitive function. *Biochem. Pharmacol.* 74, 1182-1191.

Levin, E.D., Slade, S., Johnson, M., Petro, A., Horton, K., Williams, P., Rezvani, A.H., Rose, J.E., 2008. Ketanserin, a 5-HT2 receptor antagonist, decreases nicotine self-administration in rats. *Eur. J. Pharmacol.* 600, 93-97.

Li, M.D., 2008. Identifying susceptibility loci for nicotine dependence: 2008 update based on recent genome-wide linkage analyses. *Hum. Genet.* 123, 119-131.

Li, M.D., 2010. Grand challenges and opportunities for molecular psychiatry research: a perspective. *Front Psychiatry* 1, 2.

Li, M.D., Burmeister, M., 2009. New insights into the genetics of addiction. *Nat. Rev. Genet* 10, 225-231.

Li, M.D., Cheng, R., Ma, J.Z., Swan, G.E., 2003. A meta-analysis of estimated genetic and environmental effects on smoking behavior in male and female adult twins. *Addiction* 98, 23-31.

Li, M.D., Lou, X.Y., Chen, G., Ma, J.Z., Elston, R.C., 2008. Gene-gene interactions among CHRNA4, CHRNB2, BDNF, and NTRK2 in nicotine dependence. *Biol. Psychiatry* 64, 951-957.

Lim, J.E., Papp, A., Pinsonneault, J., Sadee, W., Saffen, D., 2006. Allelic expression of serotonin transporter (SERT) mRNA in human pons: lack of correlation with the polymorphism SERTLPR. *Mol. Psychiatry* 11, 649-662.

Lou, X.Y., Chen, G.B., Yan, L., Ma, J.Z., Mangold, J.E., Zhu, J., Elston, R.C., Li, M.D., 2008. A combinatorial approach to detecting gene-gene and gene-environment interactions in family studies. *Am. J. Hum. Genet.* 83, 457-467.

Lou, X.Y., Chen, G.B., Yan, L., Ma, J.Z., Zhu, J., Elston, R.C., Li, M.D., 2007. A Generalized Combinatorial Approach for Detecting Gene-by-Gene and Gene-by-Environment Interactions with Application to Nicotine Dependence. *Am. J. Hum. Genet.* 80, 1125-1137.

Mihailescu, S., Palomero-Rivero, M., Meade-Huerta, P., Maza-Flores, A., Drucker-Colin, R., 1998. Effects of nicotine and mecamylamine on rat dorsal raphe neurons. *Eur. J. Pharmacol.* 360, 31-36.

Nelson, M.R., Kardia, S.L., Ferrell, R.E., Sing, C.F., 2001. A combinatorial partitioning method to identify multilocus genotypic partitions that predict quantitative trait variation. *Genome Res.* 11, 458-470.

Niesler, B., Flohr, T., Nothen, M.M., Fischer, C., Rietschel, M., Franzek, E., Albus, M., Propping, P., Rappold, G.A., 2001. Association between the 5' UTR variant C178T of the serotonin receptor gene HTR3A and bipolar affective disorder. *Pharmacogenetics* 11, 471-475.

Nyholt, D.R., 2004. A simple correction for multiple testing for single-nucleotide polymorphisms in linkage disequilibrium with each other. *Am. J. Hum. Genet.* 74, 765-769.

Patkar, A.A., Gopalakrishnan, R., Berrettini, W.H., Weinstein, S.P., Vergare, M.J., Leone, F.T., 2003. Differences in platelet serotonin transporter sites between African-American tobacco smokers and non-smokers. *Psychopharmacology* (Berl) 166, 221-227.

Raychaudhuri, S., 2011. Mapping rare and common causal alleles for complex human diseases. *Cell* 147, 57-69.

Ribeiro, E.B., Bettiker, R.L., Bogdanov, M., Wurtman, R.J., 1993. Effects of systemic nicotine on serotonin release in rat brain. *Brain Res.* 621, 311-318.

Ritchie, M.D., Hahn, L.W., Roodi, N., Bailey, L.R., Dupont, W.D., Parl, F.F., Moore, J.H., 2001. Multifactor-dimensionality reduction reveals high-order interactions among estrogen-metabolism genes in sporadic breast cancer. *Am. J. Hum. Genet.* 69, 138-147.

Semba, J., Wakuta, M., 2008. Chronic effect of nicotine on serotonin transporter mRNA in the raphe nucleus of rats: reversal by co-administration of bupropion. *Psychiatry Clin. Neurosci.* 62, 435-441.

Seneviratne, C., Franklin, J., Beckett, K., Ma, J.Z., Ait-Daoud, N., Payne, T.J., Johnson, B.A., Li, M.D., 2013. Association, interaction, and replication analysis of genes encoding serotonin transporter and 5-HT3 receptor subunits A and B in alcohol dependence. *Hum. Genet.* 132, 1165-1176.

Seneviratne, C., Huang, W., Ait-Daoud, N., Li, M.D., Johnson, B.A., 2009. Characterization of a functional polymorphism in the 3' UTR of SLC6A4 and its association with drinking intensity. *Alcohol Clin. Exp. Res.* 33, 332-339.

Slotkin, T.A., Seidler, F.J., 2010. Mimicking maternal smoking and pharmacotherapy of preterm labor: interactions of fetal nicotine and dexamethasone on serotonin and dopamine synaptic function in adolescence and adulthood. *Brain Res. Bull* 82, 124-134.

Staley, J.K., Krishnan-Sarin, S., Zoghbi, S., Tamagnan, G., Fujita, M., Seibyl, J.P., Maciejewski, P.K., O'Malley, S., Innis, R.B., 2001. Sex differences in [123I]beta-CIT SPECT measures of dopamine and serotonin transporter availability in healthy smokers and nonsmokers. *Synapse* 41, 275-284.

Steen, K.V., 2012. Travelling the world of gene-gene interactions. *Brief Bioinform* 13, 1-19.

Sullivan, P.F., Kendler, K.S., 1999. The genetic epidemiology of smoking. *Nicotine Tob. Res.* 1 Suppl 2, S51-57; discussion S69-70.

Swan, G.E., Carmelli, D., Rosenman, R.H., Fabsitz, R.R., Christian, J.C., 1990. Smoking and alcohol consumption in adult male twins: genetic heritability and shared environmental influences. *J. Subst. Abuse* 2, 39-50.

Swan, G.E., Hudmon, K.S., Jack, L.M., Hemberger, K., Carmelli, D., Khroyan, T.V., Ring, H.Z., Hops, H., Andrews, J.A., Tildesley, E., McBride, D., Benowitz, N., Webster, C., Wilhelmsen, K.C., Feiler, H.S., Koenig, B., Caron, L., Illes, J., Cheng, L.S., 2003. Environmental and genetic determinants of tobacco use: methodology for a multidisciplinary, longitudinal family-based investigation. *Cancer Epidemiol. Biomarkers* Prev. 12, 994-1005.

True, W.R., Heath, A.C., Scherrer, J.F., Waterman, B., Goldberg, J., Lin, N., Eisen, S.A., Lyons, M.J., Tsuang, M.T., 1997. Genetic and environmental contributions to smoking. *Addiction* 92, 1277-1287.

True, W.R., Xian, H., Scherrer, J.F., Madden, P.A., Bucholz, K.K., Heath, A.C., Eisen, S.A., Lyons, M.J., Goldberg, J., Tsuang, M., 1999. Common genetic vulnerability for nicotine and alcohol dependence in men. *Arch. Gen. Psychiatry* 56, 655-661.

Trummer, O., Koppel, H., Wascher, T.C., Grunbacher, G., Gutjahr, M., Stanger, O., Ramschak-Schwarzer, S., Boehm, B.O., Winkelmann, B.R., Marz, W., Renner, W., 2006. The serotonin transporter gene polymorphism is not associated with smoking behavior. *Pharmacogenomics J.* 6, 397-400.

Vallender, E.J., Priddy, C.M., Hakim, S., Yang, H., Chen, G.L., Miller, G.M., 2008. Functional variation in the 3' untranslated region of the serotonin transporter in human and rhesus macaque. *Genes Brain Behav.* 7, 690-697.

van der Zwaluw, C.S., Engels, R.C., 2009. Gene-environment interactions and alcohol use and dependence: current status and future challenges. *Addiction* 104, 907-914.

Walstab, J., Hammer, C., Bonisch, H., Rappold, G., Niesler, B., 2008. Naturally occurring variants in the HTR3B gene significantly alter properties of human heteromeric 5-hydroxytryptamine-3A/B receptors. *Pharmacogenet Genomics* 18, 793-802.

Walstab, J., Rappold, G., Niesler, B., 2010. 5-HT(3) receptors: role in disease and target of drugs. *Pharmacol. Ther.* 128, 146-169.

Wang, J., Li, M.D., 2010. Common and unique biological pathways associated with smoking initiation/progression, nicotine dependence, and smoking cessation. *Neuropsychopharmacology* 35, 702-719.

Ward, R.J., Lallemand, F., de Witte, P., 2009. Biochemical and neurotransmitter changes implicated in alcohol-induced brain damage in chronic or 'binge drinking' alcohol abuse. *Alcohol Alcohol* 44, 128-135.

WHO, 2008. Report on the Global Tobacco Epidemic, 2008: The MPOWER package. World Health Organization, Geneva.

Wonnacott, S., 1997. Presynaptic nicotinic ACh receptors. *Trends Neurosci.* 20, 92-98.

Wonnacott, S., Kaiser, S., Mogg, A., Soliakov, L., Jones, I.W., 2000. Presynaptic nicotinic receptors modulating dopamine release in the rat striatum. *Eur. J. Pharmacol.* 393, 51-58.

Yamada, K., Hattori, E., Iwayama, Y., Ohnishi, T., Ohba, H., Toyota, T., Takao, H., Minabe, Y., Nakatani, N., Higuchi, T., Detera-Wadleigh, S.D., Yoshikawa, T., 2006. Distinguishable haplotype blocks in the HTR3A and HTR3B region in the Japanese reveal evidence of association of HTR3B with female major depression. *Biol. Psychiatry* 60, 192-201.

Yang, Z., Seneviratne, C., Wang, S., Ma, J.Z., Payne, T.J., Wang, J., Li, M.D., 2013. Serotonin transporter and receptor genes significantly impact nicotine dependence through genetic interactions in both European American and African American smokers. *Drug Alcohol Depend* 129, 217-225.

Zuk, O., Hechter, E., Sunyaev, S.R., Lander, E.S., 2012. The mystery of missing heritability: Genetic interactions create phantom heritability. *Proc. Natl. Acad. Sci. USA* 109, 1193-1198.

In: New Developments in Serotonin Research
Editor: Ming D. Li
ISBN: 978-1-63482-220-6

Chapter 2

The Serotonergic Network and Sudden Infant Death Syndrome

Siri Hauge Opdal
Norwegian Institute of Public Health, Division of Forensic Medicine, Oslo, Norway, and Department of Pathology, Oslo University Hospital, Oslo, Norway

Abstract

Sudden infant death syndrome (SIDS) is defined as the sudden death of an infant less than one year of age, with onset of the fatal episode apparently occurring during sleep, which remains unexplained after the performance of a complete autopsy, death scene investigation and a review of the clinical history. A subset of SIDS is proposed to result from abnormalities in serotonin and related neurotransmitters in regions of the lower brainstem. This may result in failure of protective homeostatic responses to life-threatening challenges during sleep, thereby leading to SIDS when the infant at the same time is in a critical developmental stage. Several studies indicate that there is a genetic predisposition to SIDS. A number of areas have been investigated, and genes involved in the serotonergic system and brain function and development have emerged as some of the most important. The most common investigated gene in the serotonergic network is the gene encoding the serotonin transporter (5-HTT), but also genes encoding serotonergic receptors and the monoamine oxidase A (MAOA) gene, as well as several other genes encoding components involved in the serotonergic network, have been investigated with regard to SIDS. This chapter will review these findings, and also try to point out the direction for further research on this topic.

Keywords: genetics, serotonin, SIDS, sudden infant death

Introduction

Sudden infant death syndrome (SIDS) is defined as "the sudden unexpected death of an infant less than one year of age, with onset of the fatal episode apparently occurring during sleep, that remains unexplained after a thorough investigation, including performance of a complete autopsy and review of the circumstances of death and the clinical history" [1]. The syndrome is multifactorial and influenced both by developmental, environmental, and biological risk factors [2, 3]. The environmental risk factors, or trigger events, are the best known, and the hazards of prone sleeping, parental smoking and overheating are now well documented [4-7]. The biological risk factors include a vulnerable developmental stage of the central nervous system and the immune system, as well as conditions such as brainstem astrogliosis and neurochemical imbalances in the medullary serotonergic network [8-10]. In addition, it becomes more and more clear that infants who succumb to SIDS most likely have an underlying genetic predisposition, consisting of polymorphisms that might predispose infants to death in critical situations [11-13]. Genes involved in the regulation of the immune system, cardiac function, the serotonergic network and brain function and development have so far emerged as the most important with regard to sudden unexpected infant death. The purpose of this chapter is to survey the current knowledge on the genetics of the serotonergic network with regard to SIDS.

A subset of SIDS most likely results from abnormalities in serotonin (5-HT; 5-hydroxytryptamine) and related neurotransmitters in regions of the lower brainstem that are important for protective responses to life-threatening challenges [10]. Findings in medullae from SIDS cases indicate abnormal 5-HT neuron firing, synthesis, release and clearance [9, 14, 15]. The elevated levels of hypoxanthine reported in infants dying of SIDS may indicate episodes of hypoxia prior to death, possibly because of ineffective autoresuscitation due to a dysregulated 5-HT system [9, 16, 17]. Defective gasping, reported in some SIDS cases, may also be due to a dysregulated 5-HT system [18, 19]. A neurochemical imbalance in the serotonergic network could put an infant at risk of sudden death during sleep, when the infant at the same time is in a critical period in the development of autonomic and central control. A defective modulation and co-ordination of this system might be critical when the infant for instance is in a prone sleeping position, initiating a chain of events ultimately leading to sudden death.

Serotonin Transporter

The serotonergic network is under bottleneck control by the serotonin transporter (5-HTT). The gene encoding this protein is located on chromosome 17, and functional polymorphisms in the promoter, intron 2 and the 3' untranslated region have been described [20-23]. The 5-HTT promoter variant number of tandem repeat (VNTR), resulting from a 44 bp deletion/insertion, gives rise to two allelic variants; S (short) and L (long) [21]. It has been shown that the L allele gives rise to a more effective promoter, resulting in enhanced expression of the gene, and that cells with the L/L genotype produce higher levels of 5-HTT mRNA than cells with S/L or S/S genotypes [24, 25]. The S and L alleles may be further subdivided by an A/G single nucleotide polymorphism (SNP) (rs25531) situated within the promoter VNTR, and it has been reported that most likely only the A variant of the L allele (L_A) is associated with high 5-HTT mRNA levels [26]. This higher 5-HTT mRNA level may result in a more effective reuptake of serotonin, and thus a lower serotonin level at the synapses, which may contribute to the dysregulation of the serotonergic network seen in SIDS. The 5-HTT promoter polymorphism has been investigated in several SIDS populations [14, 27-33] (table 1). The first study investigating this gene in cases of SIDS was by Narita et al, who investigated 27 Japanese SIDS cases and found an association between the L allele and SIDS (27). Most of the subsequent studies report an association between SIDS and the L/L genotype or the L-allele, even though several of these studies also includes relatively few SIDS cases [27, 28, 30]. Two studies that include 145 and 179 SIDS cases respectively do not find an association [31, 32], while a study of 163 Norwegian SIDS cases found a tendency to an association with the L allele [29].

Table 1. Survey of genes involved in the serotonergic network investigated in SIDS

Gene	Variant	Number of SIDS cases	Number of controls	Findings	Comment	Reference
5-HTT	VNTR in promoter	27 Japanese	115	L-allele and L/L genotype associated with SIDS	genotype distribution p=0.009, allele distribution p=0.006	[27]
		43 African-American 44 Caucasian	43 44	L-allele associated with SIDS in total dataset and Caucasians	L/L genotype total p=0.048, L-allele total p=0.005, L-allele Caucasians p=0.023	[28]
		31 mixed ethnicity	10	no association		[14]
		20 Caucasian	150	L/L genotype and L-allele associated with SIDS	L/L genotype p=0.001, L-allele p<0.001	[30]
		163 Caucasian	243	L-allele associated with SIDS	p=0.05	[29]
		28 Caucasian	17	L-allele associated with SIDS	p=0.012 [not included in the paper but may be calculated from table 3]	[42]
		145 Caucasian	58	no association		[31]
		179 mixed ethnicity	139	no association		[32]
	VNTR in intron 2	43 African-American 46 Caucasian	43 46	12/12 genotype and 12-allele associated with SIDS in African-American	12/12 genotype p=0.03, 12-allele p=0.018	[35]
		20 Caucasian	150	no association		[30]
		163 Caucasian	243	no association		[29]

Gene	Variant	Number of SIDS cases	Number of controls	Findings	Comment	Reference
		145 Caucasian	58	no association		[31]
		147 mixed ethnicity	122	10/10 genotype possibly associated with SIDS [p<0.02]	According to the authors this may be due to bias in the control group	[36]
	rs2020939	195 Caucasian	846	no association		[37]
	rs25531	193 Caucasian	335	no association		[33]
	rs3813034	46 Caucasian 46 African-American	46 46	no association		[38]
MAOA	VNTR in promoter	43 Caucasian	41	no association		[46]
		156 Caucasian	260	2 and 3 allele associated with male SIDS	p=0.009, allele distribution in male	[48]
		142 Caucasian	280	3.5 and 4 allele associated with female SIDS	p=0.038	[49]
		128 Caucasian	147	no association		[51]
		139 mixed ethnicity	92	no association		[36]
		171 Caucasian	335	no association		[33]
		261 mixed ethnicity	93	no association		[50]
	12 SNPs	261 mixed ethnicity	93	Three SNPs possibly associated with SIDS	rs201785423 p=0.009 rs77698881 p<0.001 rs5905702 p=0.04	[50]
HTR1A	Coding sequence	49 African-American 47 Caucasian	49 47	no association		[56]

Table 1. (Continued)

Gene	Variant	Number of SIDS cases	Number of controls	Findings	Comment	Reference
HTR2A	All exons	49 African-American 47 Caucasian	49 47	no association		[60]
PHOX2B	All exons	23 Japanese	0	no association		[69]
		45 African-American 46 Caucasian	45 46	exon 3 polymorphisms possibly associated with SIDS	Increased total number of SIDS cases with one of the gene variants T495G, T536A, C552T, C642T, A726G, G750A, A762C, C870A [p=0.05]	[70]
	Exon 3 repeat	46 African-American 46 Caucasian	46 46	no association		[71]
		195 Caucasian	846	polyalanine repeat in exon 3 associated with SIDS	$p<0.001$	[37]
TPH2	rs1463	20 Caucasian	150	no association		[30]
FEV	Entire gene	49 African-American 47 Caucasian	49 47	IVS2-191_190insA associated with African-American SIDS	$p=0.01$	[65]
		78 mixed ethnicity	355	no association		[66]

Table 2. Survey of genotype and allele frequencies in merged SIDS populations

		Mixed ethnicity		p-value	Caucasian		p-value	Reference
Gene	Genotype / allele	SIDS n [%]	Controls n [%]		SIDS n [%]	Controls n [%]		
5-HTT VNTR promoter	L/L all other	235 [34.7] 442 [65.3]	283 [25.3] 835 [74.7]	<0.001	172 [34.8] 322 [65.2]	184 [25.8] 528 [75.2]	0.001	[14, 27-32, 42]
	L S	781 [57.7] 573 [42.3]	1088 [48.7] 1148 [51.3]	<0.001	584 [59.1] 404 [40.9]	729 [51.2] 695 [48.8]	<0.001	
5-HTT VNTR intron 2	12/12 all other	228 [40.4] 337 [59.6]	283 [42.7] 380 [57.3]	0.41	164 [36.7] 283 [63.3]	278 [40.9] 402 [59.1]	0.16	[29-31, 35, 36]
	12 all other	714 [63.2] 416 [36.8]	865 [65.2] 461 [34.8]	0.29	541 [60.5] 353 [39.5]	869 [63.9] 491 [36.1]	0.11	
MAOA VNTR promoter*	2 + 3 3.5 + 4 + 5	184 [38.7] 292 [61.3]	253 [32.5] 525 [67.5]	0.027	173 [39.8] 262 [60.2]	245 [32.2] 516 [67.8]	0.008	[33, 36, 46, 48, 49, 51]

* Males only.

However, when investigating the 5-HTT promoter more closely by also including the rs25531 polymorphism and investigating larger SIDS and control populations (193 SIDS cases and 335 controls), this tendency was no longer present [33]. It was found that 26.9% of the SIDS cases had the high producing L_A/L_A genotype, compared to 25% of the controls (p=0.65) [33].

If merging studies of the 5-HTT promoter VNTR in SIDS a total of 677 SIDS cases and 1118 controls have been investigated (table 2). Of these, 34.7% of the mixed SIDS population and 34.8% of the Caucasian SIDS cases had the L/L genotype, compared to 25.3% and 25.8% of the controls (p<0.001 and p=0.001, respectively) (table 2).

In intron 2, a 17 bp VNTR gives rise to three allelic variants, denoted 9, 10 and 12, and it has been shown that the 12 allele has stronger enhancer-like properties than the other alleles, and thereby results in a higher expression of the gene [20, 34]. Also this VNTR have been investigated in several SIDS populations [29-31, 35, 36] (table 1). One study that reports an association with SIDS is by Weese-Mayer et al, who found an association with the 12-allele and the 12/12 genotype in African-American SIDS [35]. The study by Paterson found an association between the 10/10 genotype and SIDS, but concludes that this is most likely not a true association, but rather due to a bias in the controls groups [36]. A Dutch study investigated rs2020939 located in the intron region of the 5-HTT gene, but it was not found any association to SIDS for this SNP [37].

In total have the intron 2 VNTR been investigated in 565 SIDS cases and 663 controls with mixed ethnicity (table 2). When merging the studies it was however not found any association between the 12/12 genotype and SIDS, neither for the whole group nor for Caucasians separately (table 2).

A few studies have investigated the combined genotype of these two VNTRs in cases of SIDS [29-31, 35]. As the long alleles of both VNTRs are associated with increased expression of 5-HTT transporters in various brain regions the synaptic serotonin levels would be expected to be lower in those infants with an L- or 12-allele, and perhaps lowest in those with both variants. An association between L/L-12/12 and SIDS has been reported in two small African-American and Italian SIDS cohorts [30, 35], this was however not confirmed in larger studies of Caucasian SIDS [29, 31].

An additional polymorphism in the 5-HTT gene investigated in SIDS is a SNP, rs3813034, situated in the untranslated region in the 3' end of the gene [22]. This SNP is located within a putative polyadenylation signal for one of the commonly used polyadenylation sites for the 5-HTT mRNA, and is therefore expected to be a useful tool for assessing the contribution of genetic variation in the serotonin transporter gene to disease. The SNP has been investigated in the same Caucasian and African-American SIDS cases as the other polymorphism in the 5-HTT gene [38], but it was not found any association to SIDS.

The investigated 5-HTT polymorphisms are hypothesized to increase the risk of SIDS by altering 5-HT neurotransmission and thereby contribute to the medullary 5-HT defects observed in SIDS. The association between genotype and brain stem pathology has been investigated in two small SIDS populations [36, 39]. The 5-HTT binding density was found to be lower in in SIDS cases with the S/L genotype of the 5-HTT promoter VNTR, and higher in the SIDS cases with the 10/10 genotype of the intron 2 VNTR. The 10/10 genotype was also associated with marginally significant higher 5-HT_{1A} receptor binding. It was not found any significant association s between the genotype and 5-HT levels or TPH2 levels with genotype for any of the 5-HTT gene polymorphisms [36]. Casale et al. performed a study where they used real-time PCR and investigated the gene expression of 5-HTT in 10 SIDS cases and one control, all previously genotyped to be S/S in the promoter VNTR [39]. In nine cases the level of 5-HTT expression was comparable to those seen in the control case, while in one case there was a reduction in the expression of the gene. The authors conclude that it is presumable that, despite the presence of the same S/S genotype, the different genetic background could influence the transcript stability and that the polymorphic variant of the 5-HTT gene could respond differently to the external environmental stimuli [39]. Taken together, these observations suggests that it is unlikely that an increased SIDS risk conferred by these polymorphisms involves a simple mechanism whereby they alter 5-HTT expression or function and lead to reduced synaptic 5-HT.

Maternal smoking is a well-known risk factor for SIDS [4, 6, 40]. The effect of nicotine prenatal exposure on serotonin transporter expression have been examined in rat brain, and it was shown significant differences in 5-HT turnover in the midbrain and pons-medulla between rats treated with nicotine injections compared with controls [41, 42]. A study by Lavezzi et al. investigated the correlation between 5-HTT genotype, maternal smoking and brainstem pathology in 28 SIDS cases and 17 controls (11 infants and 6 fetal deaths, all with known causes of death) [42]. The study revealed a correlation between the L-allele, hypoplasia of the raphe nuclei, and maternal smoking during pregnancy in both SIDS and sudden fetal death, and the authors suggests that this combination may result in failure in mediating vital functions, triggering a lethal mechanism in both intrauterine and postnatal death [42]. Duncan et al. investigated the effect of prenatal nicotine-exposure on the serotonergic network in baboons [43]. The key finding of this study that is relevant to SIDS is that prenatal exposure to nicotine is associated with alterations in the medullary 5-HT receptor binding in the raphe obscurus in the medullary 5-HT system, as well as alterations in nicotine receptor binding through this system.

Monoamine Oxidase A

In addition to 5-HTT is also monoamine oxidase A (MAOA) involved in regulating the monoamine transmitter levels in the central nervous system, by being responsible for the intracellular degradation of serotonin. The MAOA gene is located on the X chromosome, and a VNTR in the promoter is reported to influence the expression of the gene. Alleles with 3.5 or 4 copies of the repeat sequence are transcribed more efficiently, and may lead to a lower serotonin level at the synapses [44]. Several studies have investigated the relationship between this VNTR and SIDS [30, 33, 36, 45-51] (table 1). However, so far the results are conflicting, and both associations and no associations with SIDS has been reported (table 1).

The first study of the MAOA gene in SIDS was by Marzano and co-workers, who investigated only 20 SIDS cases, but did not find any clear differences between SIDS cases and controls [30]. In later studies they included stillbirths in their SIDS cases, and found an association between the 5-HTT/MAOA genotype L/L-4/4 and SIDS+SUID [45, 47]. However, they do not take into consideration that the MAOA gene is located on the X chromosome, and that it therefore is necessary to investigate male and female cases separately. A larger study from Klintschar et al. found that the lower expressing alleles were more frequent in male SIDS than in male controls (44% vs 26%) [48]. Another study of German SIDS cases found an increased frequency of the 4/4 genotype and the high expressing 3.5 and 4 alleles in female SIDS compared to female controls [49]. However, the value of deducing MAOA allele frequencies from genotype frequencies in female are limited, due to the imprinting of one of the X chromosomes in the heterozygous female [52, 53]. A study of 171 Norwegian SIDS cases did not find any association between this gene and SIDS [33]. It was also not found any higher frequency of the combined MAOA/5-HTT high producing genotypes 3.5-L_AL_A and 4-L_AL_A in male SIDS cases, compared to male controls [33].

As the MAOA gene is X-linked it might be that this gene is of greater importance for male than female SIDS. In total 476 male SIDS cases and 778 male controls with mixed ethnicity have been investigated with regard to the promoter polymorphism in this gene (table 2). Of these, 38.7% of the mixed SIDS population and 39.8% of the Caucasian SIDS cases had the low producing alleles 2 and 3, compared to 32.5% and 32.2% of the controls ($p=0.027$ and $p=0.008$, respectively) (table 2).

A study by Gross and coworkers has investigated 12 SNPs in the MAOA gene in 285 SIDS cases [50]. They observed statistically significant associations for three of the SNPs; rs201785423, rs7769881, and rs5905702. However, for none of these SNPs have there been published any significant correlation to disease before, and the results may be due to chance and a low number of investigated cases. Nevertheless, the study clearly indicates that the MAOA gene might be of importance as a predisposing factor for SIDS.

Serotonergic Receptors

Serotonin acts through several receptor types and subtypes. These can be divided into seven families, the largest being the 5-HT_1 family with five 5-HT receptor subtypes termed $5\text{-HT}_{1A\text{-}F}$, which are distinguished by different pharmacology and distribution in the brain [54, 55]. The 5-HT_{1A} receptor is important for the autonomic response to homeostatic stress and is densely expressed in medullary regions pertinent to cardiorespiratory regulation. 5-HT_{1A} receptor density has been found significantly reduced in multiple medullary nuclei of SIDS cases, especially in male infants [14]. Based on this, the coding region of the 5-HT_{1A} gene was sequenced in a cohort of 49 African-American and 47 Caucasian SIDS cases, and the same number of controls [56]. It was however not found any association between genetic variation and SIDS, indicating that abnormalities within the receptor gene itself most likely not are responsible for the decreased 5-HT_{1A} receptor binding density observed in these cases.

The receptor 5-HT_{2A} plays an important role in the physiologic response to hypoxia and is important for the generation of respiratory rhythm [19, 57]. Functional variants in the coding and regulatory regions of the 5-HT_{2A} gene have been linked to several psychiatric disorders involving neurological development [58]. A reduced 5-HT_{2A} receptor binding have been observed in the medulla of SIDS cases [59]. The gene has been examined in the same SIDS cases and controls as the 5-HT_{1A} receptor gene [60], and it was found a total of 21 gene variations. However, none of them showed any significant association with SIDS.

Tryptophan Hydroxylase

Tryptophan hydroxylase (TPH) catalyzes is the rate-limiting step in the synthesis of serotonin. There are two forms of THP; THP1 is mainly expressed in gut, pineal gland, spleen and thymus, while THP2 is specifically expressed in the brainstem, with particularly high expression in the serotonergic neurons of the raphe nuclei [61]. Analysis of the THP2 gene is relevant to SIDS because functional variants in this gene, resulting in altered expression or function of the enzyme, could result in a reduction of 5-HT synthesis.

The levels of brainstem serotonin and TPH2 levels have been investigated in a study including 35 SIDS cases and 5 controls [62]. It was found that the 5-HT levels were 26% lower and the TPH2 levels 22% lower in the SIDS cases than in the controls, and the authors conclude that this is consistent with a disorder of medullary 5-HT deficiency.

The only study investigating this gene sequenced a 720 bp region of exon IX of the THP2 gene in 20 SIDS Italian cases and 50 controls [30] (table 1). This area includes the known G1463A mutation, that gives an 80% loss of function in serotonin production [63]. However, neither this nor any other polymorphisms or mutations were detected, neither in the SIDS cases nor in the controls [30].

Human Fifth Ewing Variant

The human fifth Ewing variant (FEV) gene is specifically expressed in central 5-HT neurons in the brain, with a predicted role in specification and maintenance of serotonergic neuronal phenotype [64]. Because FEV specifically interacts with the promoter region of the 5-HTT gene one may hypothesize that variations in the FEV gene may be of importance with regard to SIDS.

The FEV gene has been investigated in two SIDS populations [65, 66] (table 1). The first consisted of 49 African-American SIDS, 47 Caucasian SIDS, and the same number of controls [65]. An association to SIDS was found only in the African-American cases, which had a higher frequency of both the insertion mutation IVS2-191_insA and a higher total variation. A study by Broadbelt et al. however found this mutation in 14% of a cohort of 78 African-American SIDS cases, as well as in 32% of African-American controls, and concluded that IVS2-191_insA seems to be a common, likely non-pathogenic, variant in the African-American population [66]. The mutation is located outside of the amino acid coding region and does not appear to be involved in splicing, but might nevertheless have a regulatory role alone or in combination with variations on other genes involved in the serotonergic network.

PHOX2B

The PHOX2B gene, which encodes a highly conserved homeobox domain transcription factor, is also of interest, as it is a key gene in the development of central serotonergic neurons. The expansion of a polyalanine repeat in exon 3 of the PHOX2B gene is reported to be associated with congenital central hypoventilation syndrome (CCHS), a condition where the patients do not response properly on elevated CO_2 levels in the blood [67, 68].

The PHOX2B gene has been investigated in four SIDS populations [37, 69-71] (table 1). Kijima et al. investigated the exons in the gene in 23 Japanese SIDS cases and did not find any mutations, concluding that PHOX2B gene variants is not likely to be associated with SIDS (69). Rand et al. sequenced the exons in 45 African-American and 46 Caucasian SIDS cases and 91 living infant controls [70]. They found that the genotype distribution for the intron 2 polymorphism IVS2+101A>G differed between SIDS cases and matched controls over the total dataset ($p<0.001$) and between the Caucasian SIDS cases and controls ($p=0.05$). In addition occurred eight polymorphisms in the third exon of the gene more frequently in SIDS cases than in controls. The authors concludes that these findings suggests that PHOX2B polymorphisms may confer some increased risk for SIDS, most likely in combination with other specific polymorphisms in the cascade of genes mediating autonomic nervous system development [70].

Two studies investigated only the exon 3 polyalanine repeat [37, 71]. One study included 92 SIDS cases, of which all had the normal genotype with 20 polyalanine-coding repeats on each chromosome [71]. The repeat has further been investigated in 195 Dutch SIDS cases and 846 controls, and this study found an association between the contraction of this repeat and SIDS, 4% of the SIDS cases had this gene variant, compared to 1% of the controls ($p<0.01$) [37].

Conclusion

Most likely SIDS has a heterogeneous nature, and rather than being a single disease entity it is likely to be polygenic in its pathogenesis. The genetic component of sudden infant death can be divided into two categories, either mutations that give rise to genetic disorders that can constitute the cause of death themselves, or polymorphisms that might predispose infants to death in critical situations. Medium-chain Acyl-CoA (MCAD) gene mutations and mutations giving long QT syndrome is an example of the first scenario, while 5-HT pathway gene variants may be an example of the latter.

Different studies of the genetic component in SIDS have often obtained conflicting results, and especially prominent have this been in the studies investigating the genes encoding 5-HTT and MAOA (table 1). This may among other thing be due to differences in the number of both included cases and controls, which for SIDS vary between 20 and 261 cases, for controls between 0 and 846 cases (table 1). In addition will diagnostic criteria for giving the SIDS diagnosis have impact, and also whether the controls are diseased or alive, or whether they are adults or infants. If the controls are diseased the cause of death may influence the results. It is further important that both SIDS cases and controls are from the same ethnic group, as even small amounts of population admixture can shift the results towards an association. It is further recommended to perform Hardy-Weinberg equilibrium (HWE) tests in association studies, as significant differences between the observed and expected genotype frequencies under HWE may indicate genotyping errors or population stratification and can increase the chance of a false-positive association (72, 73). One of the important issues in future research of genetics in SIDS is therefore to investigate the published genetic associations, or lack of such, in large well defined SIDS cohorts and control groups from several different ethnic groups.

When looked at separately, the studies performed so far indicates that commonly investigated polymorphisms in the serotonergic network as such most likely not are of importance with regard to SIDS. However, when merging the studies, both the VNTR in the 5-HTT promoter and the VNTR in the promoter of the MAOA gene shows differences in genotype frequency between SIDS cases and controls (table 2). Also, since a vast number of studies indicate that medullary 5-HT abnormalities contribute to the death in SIDS, it is important to investigate the genes involved in the serotonergic network in more detail. The polymorphisms investigated with regard to SIDS so far are only a few of several functional polymorphisms within these genes, and in addition should haplotypes be considered. Furthermore is the X-chromosomal location of the MAOA gene is intriguing, in light of the fact that more males than females die in SIDS. It would therefore be interesting to investigate other polymorphisms in this gene, as well as other X-chromosomal genes that might be of interest with regard to sudden unexpected infant death.

An attempt has been made to assess forensic biomarkers in SIDS, to help identify this group among other sudden unexpected deaths in infancy [74]. A study by Rognum et al. investigated 5-HT related substances in cerebrospinal fluid from 55 SIDS cases and 29 controls. It was however not any differences in the levels of the serotonergic markers 5-Hydroxyindoleacetic acid (5-HIAA), tryptophan, homovanillic acid (HVA) or tyrosine between the SIDS and non-SIDS group [74]. Regardless of this the study provide important information about post mortem monoaminergic measurements, and points to further research regarding the search for biomarkers for 5-HT medullary pathology in SIDS infants in other readily accessible tissue compartments, such as serum and platelets, or with different 5-HT related markers, such as TPH2.

Epigenetic phenomena might also substantially contribute to the molecular biology of SIDS. Epigenetic mutations are described as heritable changes in gene function or expression not explained by changes in DNA sequence. Even though a challenging task, the investigation of epigenetic changes in genes involved in the serotonergic network in cases of SIDS would be worthwhile.

For a truly in-depth and functional understanding of the underlying pathomechanisms of SIDS an animal model, in which by targeted mutagenesis the influence of specific genotypes could be measured and examined in convergence with external risk factors and developmental risk factors, may be essential. A few studies have tried this approach [75, 76]. Audero et al. used a mouse model with reversible overexpression of 5-HT_{1A} autoreceptor in order to investigate the role of serotonin homeostasis in SIDS [75]. It was found that overexpressing mice failed to activate autonomic target organs reacting to exterior stressors, and that they exhibited sporadic bradyrcardia frequently leading to death [75]. The authors concluded that deficient serotonin homeostasis could be linked to sporadic autonomic crisis and sudden death, and suggested that their model could be used as an SIDS relevant animal model. A 5-HTT knockout mouse model was recently introduced by Pavone et al., [76]. Seventy-five % of the 5-HTT negative mice died during the first week after birth, indicating that this model mouse generates a phenotype ultimately leading to sudden death and thereby reproducing the SIDS genotype.

The observations of the genetic studies described in this chapter do not provide compelling evidence that any of the 5-HT pathway genes studied to day play a major role in the pathogenesis of SIDS, at least in Caucasian populations. It is however important to remember that the serotonergic network constitutes a complex interplay between a range of proteins, enzymes and receptors, many of which not have been investigated neither at the genetic nor the molecular level. It remains also quite possible that other genes in a parallel network that is modulated by the serotonergic network will be of importance. In order to fully understand the importance of any genetic predisposition in the serotonergic network in SIDS these genes should be investigated in detail as well. Furthermore is the specific nature of the serotonergic dysfunction in SIDS still unclear, and whether there is an excess or a deficit of available 5-HT in SIDS is an open issue. Taken the critical role 5-HTT has in regulating the serotonergic network, it is of special importance to elucidate the role this transporter play with regard to SIDS.

References

[1] Krous HF, Beckwith JB, Byard RW, Rognum TO, Bajanowski T, Corey T, et al., Sudden infant death syndrome and unclassified sudden infant deaths: a definitional and diagnostic approach. *Pediatrics* 2004;114:234-8.

[2] Rognum TO, Saugstad OD. Biochemical and immunological studies in SIDS victims. Clues to understanding the death mechanism. *Acta Paediatr. Suppl.* 1993;82 Suppl 389:82-5.

[3] Filiano JJ, Kinney HC. A perspective on neuropathologic findings in victims of the sudden infant death syndrome: the triple-risk model. *Biol. Neonate* 1994;65:194-7.

[4] Schoendorf KC, Kiely JL. Relationship of sudden infant death syndrome to maternal smoking during and after pregnancy. *Pediatrics* 1992;90:905-8.

[5] Markestad T, Skadberg B, Hordvik E, Morild I, Irgens LM. Sleeping position and sudden infant death syndrome (SIDS): effect of an intervention programme to avoid prone sleeping. *Acta Paediatr* 1995;84:375-8.
[6] Arnestad M, Andersen M, Vege A, Rognum TO. Changes in the epidemiological pattern of sudden infant death syndrome in southeast Norway, 1984-1998: implications for future prevention and research. *Arch. Dis. Child* 2001;85:108-15.
[7] Blair PS, Sidebotham P, Berry PJ, Evans M, Fleming PJ. Major epidemiological changes in sudden infant death syndrome: a 20-year population-based study in the UK. *Lancet* 2006;367:314-9.
[8] Kinney HC, Burger PC, Harrell FE, Jr., Hudson RP, Jr. 'Reactive gliosis' in the medulla oblongata of victims of the sudden infant death syndrome. *Pediatrics* 1983;72:181-7.
[9] Kinney HC, Filiano JJ, White WF. Medullary serotonergic network deficiency in the sudden infant death syndrome: review of a 15-year study of a single dataset. *J. Neuropathol. Exp. Neurol.* 2001;60:228-47.
[10] Kinney HC, Richerson GB, Dymecki SM, Darnall RA, Nattie EE. The brainstem and serotonin in the sudden infant death syndrome. *Annu. Rev. Pathol.* 2009;4:517-50.
[11] Opdal SH, Rognum TO. The sudden infant death syndrome gene: does it exist? *Pediatrics* 2004;114:e506-e512.
[12] Courts C, Madea B. Genetics of the sudden infant death syndrome. *Forensic. Sci. Int.* 2010;203:25-33.
[13] Opdal SH, Rognum TO. Gene variants predisposing to SIDS: current knowledge. *Forensic Sci. Med. Pathol.* 2011;7:26-36.
[14] Paterson DS, Trachtenberg FL, Thompson EG, Belliveau RA, Beggs AH, Darnall R, et al., Multiple serotonergic brainstem abnormalities in sudden infant death syndrome. *JAMA* 2006;296:2124-32.
[15] Paterson DS, Hilaire G, Weese-Mayer DE. Medullary serotonin defects and respiratory dysfunction in sudden infant death syndrome. *Respir. Physiol. Neurobiol.* 2009;168:133-43.
[16] Rognum TO, Saugstad OD, Oyasaeter S, Olaisen B. Elevated levels of hypoxanthine in vitreous humor indicate prolonged cerebral hypoxia in victims of sudden infant death syndrome. *Pediatrics* 1988;82:615-8.
[17] Opdal SH, Rognum TO, Vege A, Saugstad OD. Hypoxanthine levels in vitreous humor: a study of influencing factors in sudden infant death syndrome. *Pediatr Res.* 1998;44:192-6.

[18] Poets CF, Meny RG, Chobanian MR, Bonofiglo RE. Gasping and other cardiorespiratory patterns during sudden infant deaths. *Pediatr Res.* 1999;45:350-4.

[19] Pena F, Ramirez JM. Endogenous activation of serotonin-2A receptors is required for respiratory rhythm generation in vitro. *J. Neurosci.* 2002;22:11055-64.

[20] Lesch KP, Balling U, Gross J, Strauss K, Wolozin BL, Murphy DL, et al., Organization of the human serotonin transporter gene. *J. Neural Transm Gen Sect* 1994;95:157-62.

[21] Heils A, Teufel A, Petri S, Stober G, Riederer P, Bengel D, et al., Allelic variation of human serotonin transporter gene expression. *J. Neurochem.* 1996;66:2621-4.

[22] Battersby S, Ogilvie AD, Blackwood DH, Shen S, Muqit MM, Muir WJ, et al., Presence of multiple functional polyadenylation signals and a single nucleotide polymorphism in the 3' untranslated region of the human serotonin transporter gene. *J. Neurochem.* 1999;72:1384-8.

[23] Wray NR, James MR, Gordon SD, Dumenil T, Ryan L, Coventry WL, et al., Accurate, Large-Scale Genotyping of 5HTTLPR and Flanking Single Nucleotide Polymorphisms in an Association Study of Depression, Anxiety, and Personality Measures. *Biol. Psychiatry* 2009;66:468-76.

[24] Lesch KP, Bengel D, Heils A, Sabol SZ, Greenberg BD, Petri S, et al., Association of anxiety-related traits with a polymorphism in the serotonin transporter gene regulatory region. *Science* 1996;274:1527-31.

[25] Heinz A, Jones DW, Mazzanti C, Goldman D, Ragan P, Hommer D, et al., A relationship between serotonin transporter genotype and in vivo protein expression and alcohol neurotoxicity. *Biol. Psychiatry* 2000;47:643-9.

[26] Wendland JR, Martin BJ, Kruse MR, Lesch KP, Murphy DL. Simultaneous genotyping of four functional loci of human SLC6A4, with a reappraisal of 5-HTTLPR and rs25531. *Mol. Psychiatry* 2006;11:224-6.

[27] Narita N, Narita M, Takashima S, Nakayama M, Nagai T, Okado N. Serotonin transporter gene variation is a risk factor for sudden infant death syndrome in the Japanese population. *Pediatrics* 2001;107:690-2.

[28] Weese-Mayer DE, Berry-Kravis EM, Maher BS, Silvestri JM, Curran ME, Marazita ML. Sudden infant death syndrome: association with a promoter polymorphism of the serotonin transporter gene. *Am. J. Med. Genet A* 2003;117A:268-74.

[29] Opdal SH, Vege A, Rognum TO. Serotonin transporter gene variation in sudden infant death syndrome. *Acta Paediatr* 2008;97:861-5.

[30] Nonnis MF, Maldini M, Filonzi L, Lavezzi AM, Parmigiani S, Magnani C, et al., Genes regulating the serotonin metabolic pathway in the brain stem and their role in the etiopathogenesis of the sudden infant death syndrome. *Genomics* 2008;91:485-91.

[31] Haas C, Braun J, Bar W, Bartsch C. No association of serotonin transporter gene variation with sudden infant death syndrome (SIDS) in Caucasians. *Leg. Med.* (Tokyo) 2009;11 Suppl 1:S210-S212.

[32] Paterson DS, Rivera KD, Broadbelt KG, Trachtenberg FL, Belliveau RA, Holm IA, et al., Lack of association of the serotonin transporter polymorphism with the sudden infant death syndrome in the San Diego Dataset. *Pediatr. Res.* 2010;68:409-13.

[33] Opdal SH, Vege A, Rognum TO. Genetic variation in the monoamine oxidase A and serotonin transporter genes in sudden infant death syndrome. *Acta Paediatr* 2014;103:393-7.

[34] Fiskerstrand CE, Lovejoy E, Gerrard L, Quinn JP. An intronic domain within the rat preprotachykinin-A gene containing a CCCT repetitive motif acts as an enhancer in differentiating embryonic stem cells. *Neurosci. Lett.* 1999;263:141-4.

[35] Weese-Mayer DE, Zhou L, Berry-Kravis EM, Maher BS, Silvestri JM, Marazita ML. Association of the serotonin transporter gene with sudden infant death syndrome: a haplotype analysis. *Am. J. Med. Genet. A* 2003;122A:238-45.

[36] Paterson DS. Serotonin gene variants are unlikely to play a significant role in the pathogenesis of the sudden infant death syndrome. *Respir Physiol. Neurobiol.* 2013;189:301-14.

[37] Liebrechts-Akkerman G, Liu F, Lao O, Ooms AH, van DK, Vermeulen M, et al., PHOX2B polyalanine repeat length is associated with sudden infant death syndrome and unclassified sudden infant death in the Dutch population. *Int. J. Legal Med.* 2014128;621-9.

[38] Maher BS, Marazita ML, Rand C, Zhou L, Berry-Kravis EM, Weese-Mayer DE. 3' UTR polymorphism of the serotonin transporter gene and sudden infant death syndrome: haplotype analysis. *Am. J. Med. Genet. A.* 2006;140:1453-7.

[39] Casale V, Oneda R, Matturri L, Lavezzi AM. Investigation of 5-HTT expression using quantitative real-time PCR in the human brain in SIDS Italian cases. *Exp. Mol. Pathol.* 2013;94:239-42.

[40] Blair PS, Fleming PJ, Bensley D, Smith I, Bacon C, Taylor E, et al., Smoking and the sudden infant death syndrome: results from 1993-5 case-control study for confidential inquiry into stillbirths and deaths in infancy. Confidential Enquiry into Stillbirths and Deaths Regional Coordinators and Researchers. *BMJ* 1996;313:195-8.
[41] Muneoka K, Ogawa T, Kamei K, Mimura Y, Kato H, Takigawa M. Nicotine exposure during pregnancy is a factor which influences serotonin transporter density in the rat brain. *Eur. J. Pharmacol.* 2001;411:279-82.
[42] Lavezzi AM, Casale V, Oneda R, Weese-Mayer DE, Matturri L. Sudden infant death syndrome and sudden intrauterine unexplained death: correlation between hypoplasia of raphe nuclei and serotonin transporter gene promoter polymorphism. *Pediatr. Res.* 2009;66:22-7.
[43] Duncan JR, Garland M, Myers MM, Fifer WP, Yang M, Kinney HC, et al., Prenatal nicotine-exposure alters fetal autonomic activity and medullary neurotransmitter receptors: implications for sudden infant death syndrome. *J. Appl. Physiol.* (1985) 2009;107:1579-90.
[44] Sabol SZ, Hu S, Hamer D. A functional polymorphism in the monoamine oxidase A gene promoter. *Hum. Genet.* 1998;103:273-9.
[45] Filonzi L, Magnani C, Lavezzi AM, Rindi G, Parmigiani S, Bevilacqua G, et al., Association of dopamine transporter and monoamine oxidase molecular polymorphisms with sudden infant death syndrome and stillbirth: new insights into the serotonin hypothesis. *Neurogenetics* 2009;10:65-72.
[46] Klintschar M, Heimbold C. Questionable association between a monoamine oxidase A promoter polymorphism and sudden infant death syndrome. *Neurogenetics* 2010;11:367-8.
[47] Filonzi L, Magnani C, Marzano FN. Confirmed association between monoamine oxidase A molecular polymorphisms and Sudden Infant Death Syndrome. *Neurogenetics* 2011;12:91-2.
[48] Klintschar M, Heimbold C. Association between a functional polymorphism in the MAOA gene and sudden infant death syndrome. *Pediatrics* 2012;129:e756-e761.
[49] Courts C, Grabmuller M, Madea B. Monoamine oxidase a gene polymorphism and the pathogenesis of sudden infant death syndrome. *J. Pediatr* 2013;163:89-93.
[50] Gross M, Bajanowski T, Vennemann M, Poetsch M. Sudden infant death syndrome (SIDS) and polymorphisms in Monoamine Oxidase A gene (MAOA): a revisit. *Int. J. Legal. Med.* 2014:128:43-9.

[51] Klintschar M, Rothamel T, Laer K. Monoamine oxidase A and sudden infant death syndrome. *J. Pediatr.* 2013;163:1533.

[52] Carrel L, Willard HF. X-inactivation profile reveals extensive variability in X-linked gene expression in females. *Nature* 2005;434:400-4.

[53] Pinsonneault JK, Papp AC, Sadee W. Allelic mRNA expression of X-linked monoamine oxidase a (MAOA) in human brain: dissection of epigenetic and genetic factors. *Hum. Mol. Genet.* 2006;15:2636-49.

[54] Richter DW, Manzke T, Wilken B, Ponimaskin E. Serotonin receptors: guardians of stable breathing. *Trends Mol. Med.* 2003;9:542-8.

[55] Pytliak M, Vargova V, Mechirova V, Felsoci M. Serotonin receptors - from molecular biology to clinical applications. *Physiol. Res.* 2011;60:15-25.

[56] Morley ME, Rand CM, Berry-Kravis EM, Zhou L, Fan W, Weese-Mayer DE. Genetic variation in the HTR1A gene and sudden infant death syndrome. *Am. J. Med. Genet. A* 2008;146:930-3.

[57] Fuller DD, Zabka AG, Baker TL, Mitchell GS. Phrenic long-term facilitation requires 5-HT receptor activation during but not following episodic hypoxia. *J. Appl. Physiol.* (1985) 2001;90:2001-6.

[58] Serretti A, Drago A, De RD. HTR2A gene variants and psychiatric disorders: a review of current literature and selection of SNPs for future studies. *Curr. Med. Chem.* 2007;14:2053-69.

[59] Ozawa Y, Okado N. Alteration of serotonergic receptors in the brain stems of human patients with respiratory disorders. *Neuropediatrics* 2002;33:142-9.

[60] Rand CM, Berry-Kravis EM, Fan W, Weese-Mayer DE. HTR2A variation and sudden infant death syndrome: a case-control analysis. *Acta Paediatr* 2009;98:58-61.

[61] Walther DJ, Bader M. A unique central tryptophan hydroxylase isoform. *Biochem. Pharmacol.* 2003;66:1673-80.

[62] Duncan JR, Paterson DS, Hoffman JM, Mokler DJ, Borenstein NS, Belliveau RA, et al., Brainstem serotonergic deficiency in sudden infant death syndrome. *JAMA* 2010;303:430-7.

[63] Zhang X, Gainetdinov RR, Beaulieu JM, Sotnikova TD, Burch LH, Williams RB, et al., Loss-of-function mutation in tryptophan hydroxylase-2 identified in unipolar major depression. *Neuron* 2005;45:11-6.

[64] Maurer P, Rorive S, de Kerchove dA, Schiffmann SN, Salmon I, de LY. The Ets transcription factor Fev is specifically expressed in the human central serotonergic neurons. *Neurosci. Lett.* 2004;357:215-8.

[65] Rand CM, Berry-Kravis EM, Zhou L, Fan W, Weese-Mayer DE. Sudden infant death syndrome: rare mutation in the serotonin system FEV gene. *Pediatr Res.* 2007;62:180-2.

[66] Broadbelt KG, Barger MA, Paterson DS, Holm IA, Haas EA, Krous HF, et al., Serotonin-related FEV gene variant in the sudden infant death syndrome is a common polymorphism in the African-American population. *Pediatr Res.* 2009;66:631-5.

[67] Amiel J, Laudier B, Attie-Bitach T, Trang H, de PL, Gener B, et al., Polyalanine expansion and frameshift mutations of the paired-like homeobox gene PHOX2B in congenital central hypoventilation syndrome. *Nat. Genet.* 2003;33:459-61.

[68] Weese-Mayer DE, Berry-Kravis EM, Ceccherini I, Rand CM. Congenital central hypoventilation syndrome (CCHS) and sudden infant death syndrome (SIDS): kindred disorders of autonomic regulation. *Respir Physiol. Neurobiol.* 2008;164:38-48.

[69] Kijima K, Sasaki A, Niki T, Umetsu K, Osawa M, Matoba R, et al., Sudden infant death syndrome is not associated with the mutation of PHOX2B gene, a major causative gene of congenital central hypoventilation syndrome. *Tohoku J. Exp. Med.* 2004;203:65-8.

[70] Rand CM, Weese-Mayer DE, Zhou L, Maher BS, Cooper ME, Marazita ML, et al., Sudden infant death syndrome: Case-control frequency differences in paired like homeobox (PHOX) 2B gene. *Am. J. Med. Genet. A* 2006;140:1687-91.

[71] Weese-Mayer DE, Berry-Kravis EM, Zhou L, Maher BS, Curran ME, Silvestri JM, et al., Sudden infant death syndrome: case-control frequency differences at genes pertinent to early autonomic nervous system embryologic development. *Pediatr. Res.* 2004;56:391-5.

[72] Ioannidis JP, Ntzani EE, Trikalinos TA, Contopoulos-Ioannidis DG. Replication validity of genetic association studies. *Nat. Genet.* 2001;29:306-9.

[73] Marchini J, Cardon LR, Phillips MS, Donnelly P. The effects of human population structure on large genetic association studies. *Nat. Genet.* 2004;36:512-7.

[74] Rognum IJ, Tran H, Haas EA, Hyland K, Paterson DS, Haynes RL, et al., Serotonin metabolites in the cerebrospinal fluid in sudden infant death syndrome. *J. Neuropathol. Exp. Neurol.* 2014;73:115-22.

[75] Audero E, Coppi E, Mlinar B, Rossetti T, Caprioli A, Banchaabouchi MA, et al., Sporadic autonomic dysregulation and death associated with excessive serotonin autoinhibition. *Science* 2008;321:130-3.

[76] Pavone LM, Spina A, Rea S, Santoro D, Mastellone V, Lombardi P, et al., Serotonin transporter gene deficiency is associated with sudden death of newborn mice through activation of TGF-beta1 signalling. *J. Mol. Cell Cardiol.* 2009;47:691-7.

In: New Developments in Serotonin Research ISBN: 978-1-63482-220-6
Editor: Ming D. Li

Chapter 3

Influence of Environmental and Social Stress Factors on Distribution of Serotonin in the Brain and Development of Killer and Self-Killer Organisms

N. G. Aleksidze[1,*] *and Tanseli Nesil*[2]
[1]Biological Institute, Ivane Javakhishvili Tbilisi State University, Tbilisi, Georgia
[2]Department of Psychiatry and Neurobehavioral Sciences, University of Virginia, Charlottesville, VA, US

Abstract

The brain plays a major role in the regulation of the emotional state in humans and animals. Environmental and social alterations can disturb the emotional state by changing the neurotransmission system in the brain. Environmental and social stressors are risk factors for the development of aggressive behavior and associated negative emotions (such as suicidal behavior). These risk factors interact with brain neurotransmitter systems in the etiology of pathological aggression and

* Corresponding author: N. G. Aleksidze, PhD; Abano Street 23, Tbilisi, 0154 Georgia. E-mail: nugzar_aleksidze@yahoo.com.

suicidal behavior. This chapter reviews some of the neurotransmitter systems that play a vital role in modulating aggressive behavior. Among various neurotransmitter systems, serotonin has a fundamental role in the pathophysiology of aggressive behavior, which is the focus of this chapter. We also discuss the effect of sex and circadian rhythm on the development of aggressive behavior in the distribution of brain neurotransmitters. This chapter reviews current findings about cognitive therapy and a serotonin-rich diet as a new non-pharmacological treatment for aggressive individuals. Finally, we discuss the effect of environmental and social stressors on human emotional disturbance with their biological interactions in the brain. It is our hope that this chapter will provide a better understanding of pathophysiology leading to aggressive behavior.

Keywords: serotonin, stress, aggression, killer rats

Introduction

Previous human and animal behavioral studies have shown how the central nervous system (CNS) is influenced by factors from the outside world that induce changes in the emotional state. Social and environmental stressors have a major role in the induction of pathological aggression and the development of killer and self-killer organisms. This psychobiological problem has been widely studied by neurochemists, neurobiologists, and psychobiologists. Common social and environmental stressors that lead to aggressive behavior are: absence of electricity, cold temperature, deficiency of food, and social isolation. These stress factors were defined as high risks for murder and suicidal behavior in former Soviet countries. In Russian Georgia, the number of murder and suicide cases exceeded 5,000 per year after the disintegration of the Soviet Union (Aleksidze 2010; Kostowski 1978; Nugzar 2008).

Literature findings (Aleksidze 2013; Ferris et al. 1997; Nishizawa et al. 1997; Schenkel 1947) suggest that the fundamental cause of aggression and murder behavior is serotonin deficiency. The relation between serotonin and aggression behavior has widely been investigated in animal models. Exposing rats to a high- or non-tryptophan- (a serotonin precursor molecule) containing diet is a common animal model to induce aggressive behavior and associated alterations in rats. It has been shown that exposing rats to a tryptophan-free diet for 4–6 days led to a 26% reduction in the brain serotonin concentration with an increase in killing behavior (Carrillo et al. 2009; Gibbons et al. 1979).

On the other hand, when the diet was supplemented with 25–100 mg of tryptophan, the killing behavior decreased as the serotonin concentration increased to baseline values (Gibbons et al. 1979). Similar trends were observed in healthy males who showed tryptophan depletion with an increase in their aggressive behavior response (Moeller et al. 1996). Aggressive behavior can be suppressed through 5-HT_{1B} receptors in animals (Ferris et al. 1997). Ferris and colleagues reported an interaction between the concentration of vasopressin/serotonin and aggressive behavior in golden hamsters (Ferris et al. 1997). In their experiments, microinjection of arginine vasopressin (AVP) into the anterior hypothalamus increased aggressive behavior. This effect was blocked by pretreatment with the selective 5-HT re-uptake inhibitor fluoxetine (Ferris et al. 1997). Serotonin and vasopressin are two important neurotransmitter systems implicated in the control of aggressive behavior. Therefore, determination of the neuromodulatory mechanisms of these systems can expand our understanding of pathological aggression and connected killer behavior.

Social isolation stress plays a critical role in the development of pathological aggression in both animals and humans. Apparently, social isolation stimulates formation of intra- and inter-specific killer behavior, and 80% of rats exposed to social isolation exhibit aggressive behavior (Carrillo et al. 2009; Miczek 1974, 1976). Studies also pointed out the effect of sex differences on aggression and killing behavior. Aggression levels were 8.5 times higher in male rats than in female rats (Kostowski 1978; Miczek 1974, 1976). Studies with human subjects showed the mean rate of serotonin synthesis in men was 40%–52% higher than in women (Nishizawa et al. 1997; Okazawa et al. 2000). Motivational analysis has revealed that aggression and killer behavior share a psychophysiological basis. Alterations in the neurotransmitter and endocrine systems contribute to expression of such behavior (Albert et al. 1986; Aleksidze 2010). A significant body of research has shown that hunger intensifies aggression. However, studies with killer rats demonstrate that such behavior also was high in satiated rats (Albert et al. 1986; Miczek 1974).

Electrical stimulation of the central-medial amygdaloid nuclei increases fight and fear reaction in animals (Davis and Whalen 2001). Aggressive outbursts may be elicited by central gray matter and dorsolateral hypothalamus stimulation. A similar phenomenon has been observed in cats with high cerebral lesions. Surgical removal of the olfactory bulbs enhances emotional tension and stimulates aggression and killing behavior in animals. Bilateral lesions in the septal area enhance intraspecific aggression in cats, dogs and rats

(Albert et al. 1986; Miczek 1976). Despite the above-described facts, the relation between the psycho-emotional state and serotonin distribution within the brain of aggressive animals remains an urgent problem in modern social psychology (Aleksidze 2010; Aleksidze 2013; Kostowski 1978).

Proceeding from these findings, we decided to study the influence of an aggressive social environment in the formation of chronic stress on killer and self-killer organisms. We also determined the effect of these phenomena on the quantitative distribution of biogenic amines (as a neurochemical correlate) in naturally aggressive and pilocarpine-induced aggressive rats. In the first series of the experiments, quantitative distribution of biogenic amines in multiple brain regions of aggressive rats was estimated. Experiments were performed on male albino rats weighing 100-150 g. Rats with an aggressive nature were selected according to Mizek's "killer-rats model" (Miczek 1976). Subchronic injection of pilocarpine at 12.5 mg/kg significantly changed rat behavior (Aleksidze 1988; Goguadze et al. 2006). Rats receiving pilocarpine were different in exploratory and stereotypical activity from rats with an aggressive nature. Chronic daily treatment with pilocarpine at doses of 12.5 or 25 mg/kg per day for three weeks induced killing behavior with an increase of 25–70% in previously non-killer rats. Pilocarpine induced aggressive behavior in a dose-dependent manner, and this effect was more pronounced under a condition of social isolation stress (Miczek 1976). Furthermore, the concentration of serotonin decreased by more than 50% in aggressive rats. The effect of pilocarpine is related to direct interaction with muscarinic acetylcholine receptors. Acetylcholine neurotransmitter, anticholinergic agents, agonists, and antagonists all enhance the development of aggressive behavior (Aleksidze 1988). The concentrations of biogenic amine and catecholamines in different brain regions of naturally aggressive and pilocarpine-induced aggressive rats are presented in Table 1. Natural and pilocarpine-induced aggressive rats showed a 1.5–2.0-fold increase in the midbrain dopamine and norepinephrine concentrations compared with their control groups. At the same time, the concentration of serotonin decreased by more than 50% in aggressive rats (Goguadze 2004). It is remarkable that serotonin synthesis is suppressed by pilocarpine, and this interaction might also amplify aggressive behavior in rats (Aleksidze 2007; Carrillo et al. 2009).

In fact, acetylcholine, anti-cholinergic agents, anti-cholinesterase, cholinergic agonists, and antagonists are involved in the neuronal mechanisms of aggressive behavior. Therefore, cholinergic neurotransmission requires more investigation to determine the aggression-related behaviors. Studies have shown that cholinergic agonists lead to high levels of fight response in the isolated animals (Aleksidze 1988; Nishizawa et al. 1997). Cholinergic drugs and pilocarpine induced a remarkable aggressive behavior response half an hour after the injection that continued for 24–48 hours. About 15–20% of animals showed the killer reaction within this time span (Aleksidze 1988; Isel et al. 1988).

Table 1. Quantitative distribution of biogenic amines (μg/g) in various regions of rat brain in naturally aggressive and pilocarpine-treated aggressive rats

Brain Region	Control	Aggression	
		Natural	Pilocarpine
Midbrain			
Dopamine	0.32±0.04	0.80±0.13	0.52±0.03†
Norepinephrine	0.48±0.02	0.68±0.21†	1.20±0.18†
Serotonin	0.50±0.05	0.35±0.06†	0.25±0.02*
Hypothalamus			
Dopamine	0.58±0.05	0.95±0.07*	0.73±0.04*
Norepinephrine	1.45±0.08	2.24±0.13†	2.83±0.02*
Serotonin	1.12±0.09	0.24±0.04*	0.45±0.06*
Hippocampus			
Dopamine	0.61±0.09	0.85±0.02†	0.74±0.04*
Norepinephrine	0.75±0.06	1.12±0.17†	1.18±0.04†
Serotonin	0.48±0.04	0.24±0.04†	0.28±0.06†

Note: *P<0.01 and †P<0.05 in comparison to the control, where P is the reliability coefficient.

The ratio between norepinephrine and serotonin plays an important role in integrated brain activity (Aleksidze 1988; Aleksidze 2013). As shown in Table 1, the ratio between norepinephrine and serotonin in the lateral hypothalamus increased six times after the decrease in serotonin concentrations in the naturally and pilocarpine-induced aggressive rats. Furthermore, pilocarpine-induced aggressive rats showed a 1.5–2.0-fold increase in norepinephrine concentration in the midbrain and hippocampus. The same group of rats showed a two-fold decrease in brain serotonin (Aleksidze 2013). A similar shift in the quantitative ratio between norepinephrine and serotonin was found in naturally aggressive rats, but to a lower degree.

Social isolation stress and low temperatures caused aggressive behavior in experiments in the former Soviet countries. To determine the relation between these stressors and the distribution of brain serotonin concentration, we exposed rats to social isolation, low temperature, and light deprivation for a defined time period. Experiments were performed on male albino rats weighing 100–150 g. Experimental rats were maintained in individual cages (social isolation) at the 1:23 light/dark regime and 15–18°C temperatures for 4 weeks. Control animals were kept together in the same cage under a 14:10 light/dark regimen (Aleksidze 2013; Goguadze 2004; Isel et al. 1988). Quantitative distribution of serotonin was investigated in the midbrain, hippocampus, lateral hypothalamus, and visual cortex (Aleksidze 1988; Aleksidze 2013; Goguadze 2004; Isel et al. 1988). The amounts of biogenic amines were determined by high-performance liquid chromatography with an electrochemical detector.

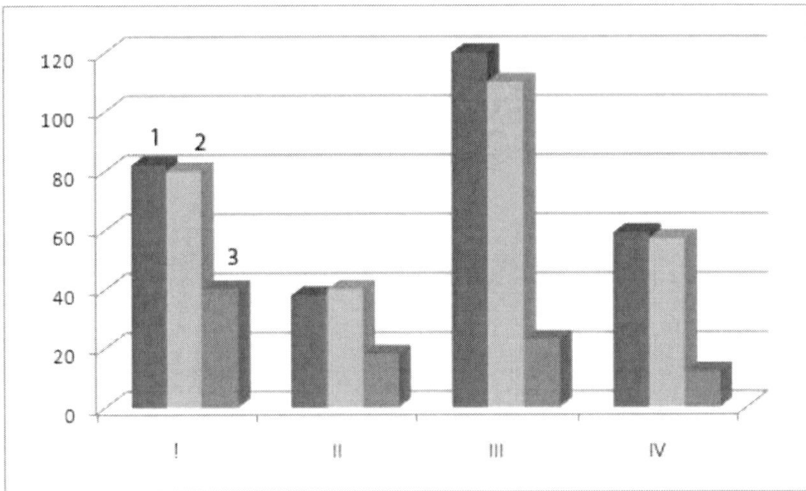

Figure 1. Effect of light deprivation combined with low temperature and social isolation on the quantitative distribution of brain serotonin (μM/g) in aggressive and naturally non-aggressive rats. I. Midbrain; II: hippocampus; III: lateral hypothalamus, IV: visual cortex. Series 1: Non-aggressive rat; Series 2: non-aggressive rat in the dark; Series 3: aggressive rat in the dark.

As shown in Figure 1, when light deprivation and social isolation were combined with low temperature, both non-aggressive and aggressive rats showed low brain serotonin concentrations. It seems that the formation of the aggressive mood and killer behavior correlates with a decrease of brain serotonin. For example, if the concentration of serotonin displayed a two-fold decrease in the midbrain and hippocampus, the same index showed a decrease of approximately five times in the lateral hippocampus and visual cortex.

Table 2. Influence of light deprivation in combination with low temperature on quantitative distribution of serotonin in various regions of brain in naturally aggressive and non-aggressive rats (M/g)

Brain Region	Non-Aggressive	Non-Aggressive in the Dark	Aggressive in the Dark
Mid-brain	0.889 ± 0.039	0.823 ± 0.083*	0.418 ± 0.066*
Hippocampus	0.465 ± 0.041	0.432 ± 0.068*	0.195 ± 0.049*
Lateral Hippocampus	1.212 ± 0.081	1.140 ± 0.115†	0.302 ± 0.088†
Visual Cortex	0.598 ± 0.028	0.581 ± 0.089*	0.143 ± 0.056*

Note: * P<0.01 and †P<0.05 in comparison to the control, where P is the reliability coefficient.

It is not hard to imagine the gravity of a prisoner's psychological condition in an aggressive and isolated environment (such as a jail or prison). In order to facilitate integration into society after release, it is necessary to assist former prisoners by providing psychopharmacological treatment after the expiration of their sentence (Aleksidze 1988).

Alterations in monoamine oxidase (MAO) activity have been identified during the chronic stress (Aleksidze 2013; Goguadze 2004). The initial stage of aggression development is associated with a significant increase in MAO activity in the frontal lobe, hypothalamus, and parietal regions. After the development of killer behavior in rats, the activity of MAO decreased by 30–40% in all brain regions (Goguadze 2004).

It is worth noting that after chronic stress, the percentage of aggressive behavior in rats was estimated as 60%, and the other 40% of the rats showed killer behavior. The background of physiological stress against the development of aggressive and killer animals is biochemically unidirectional.

It is also necessary to underline the impact of circadian rhythms in the development of aggressive behavior. Using different types and levels of light can induce impairment in circadian rhythm and results in aggressive behavior in rats. Analysis showed that light deprivation combined with low temperature, and isolation stress induced low serotonin levels in aggressive rats. It is known that a low level of serotonin is the index of abnormal aggression behavior (Aleksidze 2013; Goguadze 2004).

Clinical studies have implicated that circadian rhythms can modulate the human erythrocyte membrane redox system by melatonin. Chronic stress can impair the circadian rhythm through deregulation of the serotonergic system. The following series of experiments was conducted with the goal of determining how chronic stress is reflected in the structural organization of the erythrocyte membrane proteins. These data can be applied in the diagnosis of the human and animal emotional state. At this point, in order to determine the level of aggressive behavior, we developed a test to detect the amount of erythrocyte membrane protein in stressed rats by using the lectin-binding technique.

Lectin-binding activity was estimated as the minimum concentration of lectins that causes agglutination of trypsin-treated rat's erythrocytes (Aleksidze 2012, 2013). Specific lectins with different affinities for terminal carbohydrates of the erythrocyte membrane were selected for experiments: VAL = *Viscum album* [mistletoe] (hapten Gal, NANA); PNA = *Arachis hypogaea* [peanut] (hapten β-D-Gal, Gal-β-(1,4-glc); PSE = *Pisum sativum* [common pea] (hapten Man, Glc); VGA = *Triticum aestivum* [wheat] (hapten GlcNac, NANA); SN = *Sambucus nigra* (hapten NANA).

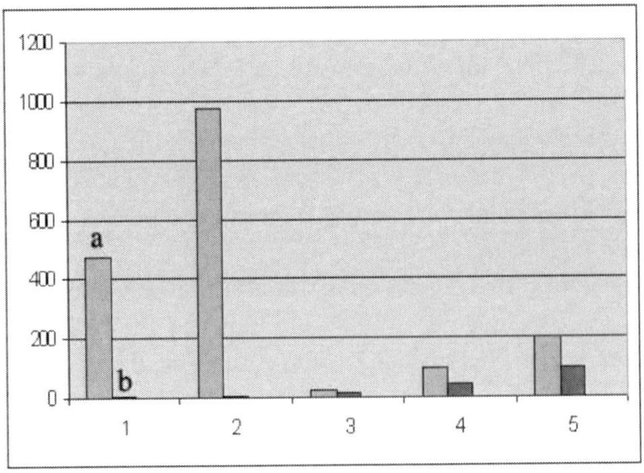

Figure 2. Lectin binding with terminal carbohydrates of trypsin-treated erythrocyte membrane of normal (a) and stressed (b) rats in accordance with lectin specificity. Y axis shows the binding activity of lectins to rat erythrocytes calculated as μM/g × 10. X axis presents the lectin types: 1) VAL; 2) PNA; 3) PSA; 4) WGA; and 5) SNA.

Figure 2 demonstrates the difference between the control and stressed rats in the lectin-binding ability of erythrocyte membrane proteins (Aleksidze

1988). In the stressed rats, erythrocyte agglutination and lectin-binding ability were lower than in the control rats. Among the erythrocyte membrane proteins, the β-D-galactose-binding lectin PNA induced low agglutination of trypsin-treated erythrocytes in the stressed rats. Specific lectin-binding of PNA also showed a significant decrease, more than 128 times for SNA and 2 times for WGA. No significant difference was found in the experiments with PSA (Aleksidze 1988).

Later, studies were carried out of lectin-binding ability of PNA to the ghost glycoconjugates of erythrocyte membranes in the blood samples of the rats. Erythrocyte membranes with glycoconjugate receptors were isolated from normal and stressed rats by the treatment of the ghost glycoconjugates with 0.1% triton X-100 phosphate-buffered saline. Analytical electrophoresis of the protein fraction 1 was carried out using the Hoefer Scientific Instruments SE-200 device in the gradient of polyacrylamide gel (10%–25%) according to Laemli's method (Laemmli 1970).

The protein fraction isolated with Triton X-100 solution demonstrated the lectin-binding and agglutinating activity. The results indicate that the glycoconjugate extracted with Triton X-100 plays an important role in the lectin-binding receptor activity. It is clear that erythrocyte membranes are modified in the stressed rats, which shows a significant decrease in the binding ability of D-galactose-specific lectin PNA.

These findings evoke the question whether the reduction of the amount of lectin-binding protein receptors or reorganization of terminal carbohydrates–glycoproteins on the erythrocyte membrane is responsible for the inhibition of lectin-binding ability in the stressed rats. To answer this question, the protein fraction was isolated by electrophoresis from the stressed and control rats (Figure 3).

The protein fraction consisted of three subunits with molecular weights of 25, 37, and 50 kDa. Schiff reactions showed that these proteins belong to the glycoprotein family. However, we failed to detect any quantitative changes of individual protein fractions. These results indicate that stress decreases the lectin-binding capacity of erythrocytes membranes. This suggests rearrangement of carbohydrate terminals of lectin-binding proteins in the erythrocyte membranes in the stressed rats.

When considered together, these results indicate that behavioral circadian rhythms are disrupted by chronic stress exposure. Distribution difference in carbohydrate terminals of the erythrocyte membrane leads to development of pathological aggressive behavior in rats. In the future, this dynamic interaction could be used in the diagnosis of aggressive behavior.

It is necessary to note that the immune complex concentration in erythrocytes was different in stressed and control rats after the polyethylene glycol antigen injections. Particularly, the binding of C3b to erythrocyte membrane CR1 receptors increases immune activity. These findings suggest that stress evokes visible changes in both the erythrocyte membrane and the immune system (Goguadze et al. 2006).

In fact, behavioral and neuroendocrine systems change with the influence of nitric oxide (Chiavegatto and Nelson 2003). This compound has an array of functions, including regulation of synaptic plasticity, the sleep–wake cycle, and hormone secretion (Chiavegatto and Nelson 2003).

Chiavegatto and colleagues showed that inhibition of the neuronal NO synthesis enzyme (NO synthase) increased the aggressive behavior of male mice. Nitric oxide synthase-knockout mice showed low brain serotonin concentrations as a result of the hypoactivity of $5HT_{1A}$ and $5HT_{1B}$ receptors. These mice also exhibited high levels of aggressive behavior under normal and isolation stress conditions. Increased aggression in knockout mice was suppressed by the injection of the serotonin precursor tryptophan and agonists. This suggests that serotonin and NO are potential targets in the treatment of aggressive behavior (Chiavegatto and Nelson 2003).

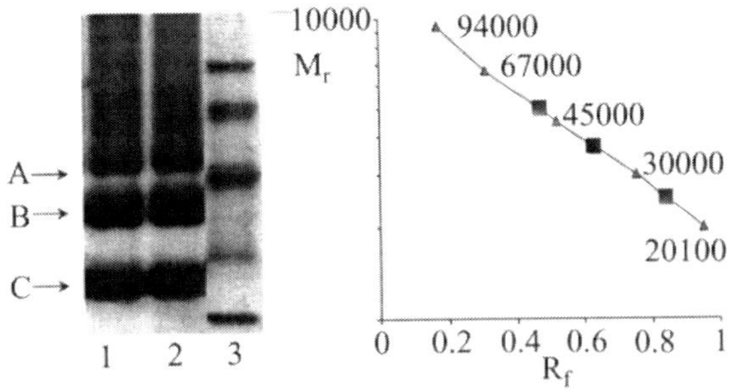

Figure 3. Electrophoresis of protein fraction rat erythrocyte ghosts extracted with 0.1% triton X-100 in dissociating conditions in the gradient of the polyacrylamide gel (10%–25%). 1. Control rats; 2. Stress-experienced rats; 3. Molecular mass of standard proteins, in descending order: 1) phosphorylase b (94 kDa); 2) albumin (67 kDa); 3) ovalbumin (45 kDa); 4) carboanhydrase (30 kDa); 5) inhibitor of trypsin (20.1 kDa); A, B, C = glycoproteins.

Suicidal behavior can correlate with the serotonergic system. The low level of serotonin-induced aggression has a strong relation to suicidal behavior. Suicide with a firearm constitutes 60% of cases, although there is a sex difference within this type of behavior. Firearm suicide has been found in 80% of the male population that exhibits suicidal behavior. The prevalence of suicide behavior in aggressive social environments is increased by four times in the male population. These results suggest that males are more vulnerable to environmental social stressors; therefore, they show high aggressive and suicide behavior compared with females. Sex differences in the brain serotonin concentration may cause the difference in this behavior.

There is a negative correlation between suicidal behavior and serotonin concentrations in the brain. Patients with suicidal behavior showed low concentrations of the molecule in the cerebrospinal fluid and various brain regions. Similarly, low concentrations of serotonin are observed in aggressive individuals who are under the influence of alcohol and in children with an aggressive attitude toward animals. Consistent with the human study findings, Aleksidze and colleagues demonstrated that the loss of dominant status in the tribe by a monkey evokes aggressive behavior with a significant reduction of the serotonin concentrations in the brain and blood (Aleksidze 2010).

Psychopharmacotherapy has an important role in the treatment of murder and suicide attempts in humans. Clinical studies have reported a significant reduction in the murder and suicide rates in patients subjected to cognitive therapies. These patients showed 50% fewer attempts at suicidal and murderous behavior (Aleksidze 2010).

It is necessary to indicate that seasonal fluctuations have a potential role in suicide and aggressive behavior and are considered a risk factor. For instance, sunlight is a bioclimatic factor that may act as a trigger of suicidal behavior (Petridou et al. 2002). Aggressive action depends on the time of the year and is particularly manifested throughout the high-temperature periods (Figure 4). This finding is in complete agreement with the results shown by Anderson (Anderson 1989).

It is important to underline that aggression-related behaviors increase during the warm months (i.e., from June to September) (Anderson 1989). We think that the decrease in such behaviors in the winter months is a result of an abnormal increase in serotonin synthesis under the influence of the winter sun. Thus, the mood of individuals with aggression improves, and psycho-emotional tension decreases. This leads to a decline in aggression-related behaviors during the winter months. It appears that suicide and pathological aggressive behavior are developed by changes in circadian rhythms as a result of the effect of seasonal fluctuations on the brain neurotransmitter systems. It has been well established that serotonin concentrations, serotonin release, and synthesis-related enzymes are affected by circadian rhythms (Barassin et al.

2002; Weiner et al. 1992). Seasonal fluctuations act as an environmental factor on circadian rhythms and disturb the emotional state by changing the serotonin concentrations in the brain.

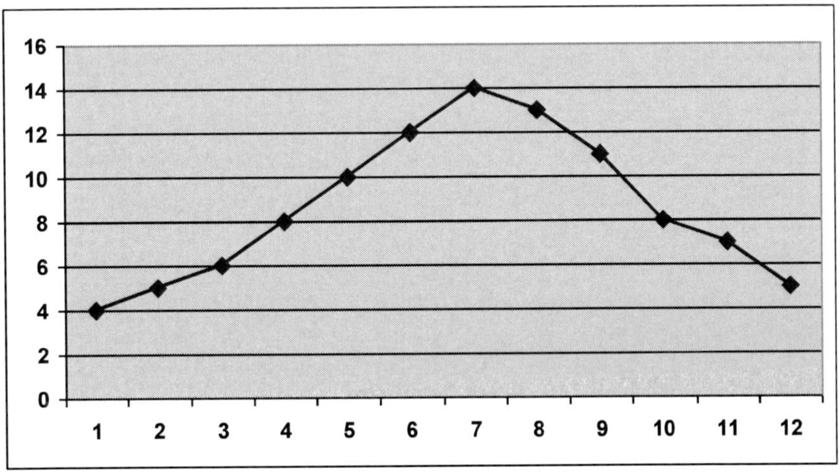

Figure 4. Time course of manifestation of violent incidents during the year. Months are shown on the X axis and cases of aggressive behavior as a percent distribution on the Y axis.

Environmental factors meet the biological and psychological factors of individuals' emotional state and influence brain function. Nutrition and exercise have a significant impact on manipulation of the serotonergic system and aggressive behavior. In order to prevent aggression, it is desirable to use non-pharmacological methods to increase serotonin concentrations; such methods include: 1) consuming foods rich in carbohydrates (wheat, potato, vegetables, and rice); 2) consuming tryptophan-rich proteins (cottage cheese, low fat milk, fish, meat, chicken, turkey, bananas, dried dates, peanuts, beans, soybean); or 3) increasing physical activity (therapeutic massage, yoga exercises). Listening to classical music and reading literature positively affect brain serotonin concentrations. On the other hand, it is necessary to indicate that a high concentration of serotonin has adverse effects on sexual activity and aggressive behavior (Nishizawa et al. 1997). Therefore, individuals taking antidepressants such as selective serotonin reuptake inhibitors (SSRIs) often have various sexual problems and become less aggressive because of their higher concentration of serotonin.

Conclusion

Social and environmental stress are high-risk factors for the development of aggressive behavior. Studies reviewed in this chapter provide numerous findings suggesting that serotonin regulation and extracellular concentrations of serotonin play a fundamental role in the development of aggressive behavior. Studies with aggressive rats provide evidence that dopamine, norepinephrine, acetylcholine, and NO are also affected by environmental and social stress factors and contribute to the development of aggressive behavior. Aggression has been linked to other negative emotional states such as suicidal behavior. Aggression and suicidal behavior are both affected by seasonal fluctuations through changes in the neurotransmitter systems. Circadian rhythm and associated molecules strongly respond to seasonal fluctuations. We also discussed a potential effective test for detection of aggressive behavior in humans and animals. Findings from these studies indicate that the lectin-binding ability of erythrocyte membranes was reduced in stressed rats. However, there are still discrepancies that need to be elucidated by future research. Sex differences in the brain neurotransmitters and aggressive behavior show an apparent need to determine the biological components of aggressive behavior. Non-pharmacological treatments of aggressive and suicidal behavior were briefly discussed. Recent findings indicate that the manipulation of the serotonergic system by nutrition and cognitive therapy will help people who struggle to control aggressive behavior and related mood disorders.

References

Albert DJ, Walsh ML, Zalys C, Dyson E (1986) Defensive aggression toward an experimenter: no differences between males and females following septal, medial accumbens, or medial hypothalamic lesions in rats. *Physiol Behav* 38: 11-4.

Aleksidze N, Alexsidze, G. (2010) *Foundations of Ecological Biochemistry.* Universal.

Aleksidze NG (1988) *Molecular and Cellular Mechanisms of Brain Integrative Activity.* Tiblisi State University.

Aleksidze NG (2007) Chronic Stress and Pathological Aggression as Premise for Killer Rats Formation. B*ull. Georg.Nat. Acad.Sci* 175.

Aleksidze NG (2012) The involvement of lectin pH3(40) in the thermoregulation of the rat brain. *Neurochemical Journal* 6: 265-267. doi: Doi 10.1134/S1819712412040022.

Aleksidze NG (2013) The involvement of serotonin and sex hormones in the generation of "killer rats". *Neurochemical Journal* 7: 266-269. doi: Doi 10.1134/S1819712413040028.

Anderson CA (1989) Temperature and aggression: ubiquitous effects of heat on occurrence of human violence. *Psychol Bull* 106: 74-96.

Barassin S, Raison S, Saboureau M, Bienvenu C, Maitre M, Malan A, Pevet P (2002) Circadian tryptophan hydroxylase levels and serotonin release in the suprachiasmatic nucleus of the rat. *European Journal of Neuroscience* 15: 833-840. doi: DOI 10.1046/j.1460-9568.2002.01928.x.

Carrillo M, Ricci LA, Coppersmith GA, Melloni RH, Jr. (2009) The effect of increased serotonergic neurotransmission on aggression: a critical meta-analytical review of preclinical studies. *Psychopharmacology (Berl)* 205: 349-68. doi: 10.1007/s00213-009-1543-2.

Chiavegatto S, Nelson RJ (2003) Interaction of nitric oxide and serotonin in aggressive behavior. *Horm Behav* 44: 233-41.

Davis M, Whalen PJ (2001) The amygdala: vigilance and emotion. *Mol Psychiatry* 6: 13-34.

Ferris CF, Melloni RH, Jr., Koppel G, Perry KW, Fuller RW, Delville Y (1997) Vasopressin/serotonin interactions in the anterior hypothalamus control aggressive behavior in golden hamsters. *J Neurosci* 17: 4331-40.

Gibbons JL, Barr GA, Bridger WH, Leibowitz SF (1979) Manipulations of Dietary Tryptophan - Effects on Mouse Killing and Brain-Serotonin in the Rat. *Brain Research* 169: 139-153. doi: Doi 10.1016/0006-8993(79) 90380-9.

Goguadze Rea (2004) *Proc. Georg. Acad. Sci* 30: 363–366.

Goguadze RP, Chachua MV, Keburiya ND, Chipashvili MD, Aleksidze GY, Aleksidze NG (2006) Changes on the surface of erythrocyte membrane during chronic stress in rats. *Bull Exp Biol Med* 141: 581-3.

Isel F, Ciesielski L, Gobaille S, Molina V, Mandel P (1988) Alterations of Serotonin Neurotransmission and Inhibition of Mouse Killing Behavior .2. Effects of Selective and Reversible Monoamine-Oxidase Inhibitors of Type-A. *Pharmacology Biochemistry and Behavior* 29: 97-104. doi: Doi 10.1016/0091-3057(88)90281-X.

Kostowski W (1978) Effects of sedatives and major tranquilizers on aggressive behavior. *Mod Probl Pharmacopsychiatry* 13: 1-12.

Laemmli UK (1970) *Cleavage of structural proteins during the assembly of the head of bacteriophage* T4. Nature 227: 680-5.

Miczek KA (1974) Intraspecies Aggression in Rats - Effects of D-Amphetamine and Chlordiazepoxide. *Psychopharmacologia* 39: 275-301. doi: Doi 10.1007/Bf00422968.

Miczek KA (1976) Mouse-Killing and Motor-Activity - Effects of Chronic Delta-9-Tetrahydrocannabinol and Pilocarpine. *Psychopharmacology* (Berl) 47: 59-64. doi: Doi 10.1007/Bf00428702.

Moeller FG, Dougherty DM, Swann AC, Collins D, Davis CM, Cherek DR (1996) Tryptophan depletion and aggressive responding in healthy males. *Psychopharmacology (Berl)* 126: 97-103. doi: Doi 10.1007/Bf02246343.

Nishizawa S, Benkelfat C, Young SN, Leyton M, Mzengeza S, DeMontigny C, Blier P, Diksic M (1997) Differences between males and females in rates of serotonin synthesis in human brain. *Proceedings of the National Academy of Sciences of the United States of America* 94: 5308-5313. doi: DOI 10.1073/pnas.94.10.5308.

Nugzar A (2008) *Normal and Pathological Biochemistry for Medical Students.* Inovatsia.

Okazawa H, Leyton M, Benkelfat C, Mzengeza S, Diksic M (2000) Statistical mapping analysis of serotonin synthesis images generated in healthy volunteers using positron-emission tomography and alpha-[(11)C]methyl-L-tryptophan. *Journal of Psychiatry & Neuroscience* 25: 359-370.

Petridou E, Papadopoulos FC, Frangakis CE, Skalkidou A, Trichopoulos D (2002) A role of sunshine in the triggering of suicide. *Epidemiology* 13: 106-109. doi: Doi 10.1097/00001648-200201000-00017.

Weiner N, Clement HW, Gemsa D, Wesemann W (1992) Circadian and Seasonal Rhythms of 5-Ht Receptor Subtypes, Membrane Anisotropy and 5-Ht Release in Hippocampus and Cortex of the Rat. Neurochemistry International 21: 7-14. doi: Doi 10.1016/0197-0186(92)90062-V.

In: New Developments in Serotonin Research ISBN: 978-1-63482-220-6
Editor: Ming D. Li

Chapter 4

Recent Discoveries on Serotonin Developmental Functions: What We Learned from *Xenopus*

Michela Ori[1], Irma Nardi[1] and Stefania De Lucchini[2]
[1]Unità di Biologia Cellulare e dello Sviluppo,
Dipartimento di Biologia, Università di Pisa, Pisa, Italy
[2]Scuola Normale Superiore di Pisa, Pisa, Italy

Abstract

Serotonin (5-HT) is an ancient molecule that appeared very early in evolution and is present in different phyla. The 5-HT signaling system is widely conserved, so researches in different animal models have general relevance for all vertebrates, including humans. 5-HT is implicated in an astonishing number of biological processes and therefore attracted studies in different fields, from neurobiology to developmental biology and neuropsychiatry. 5-HT is present in the egg and in the placenta, which can synthesize it from maternally derived tryptophan and has a key role as a morphogen in several complex networks during development before it can act as a neurotransmitter. Recent advances on how serotonin signaling can influence early development come from *Xenopus*, thanks to several advantages over other model systems such as rodents. *Xenopus* studies highlight the role of 5-HT in establishing the left-right asymmetry. These findings have relevance in molecular medicine because a common class of birth defects is due to laterality errors.

Recent advances also show that 5-HT is critically required for craniofacial and eye morphogenesis. In particular, 5-HT signaling is necessary for the correct development of the first branchial arch derivatives, and to define and sustain the gene expression network necessary for the jaw and jaw joint formation, which is one of the major novelties of vertebrates. On the other hand, craniofacial disorders are a primary cause of infant mortality and have serious lifetime functional, aesthetic, and social consequences. Of all the babies born with birth defects, approximately one-third display anomalies of the head and face. For these reasons the possibility to modulate selectively the activity of the only serotonin receptor expressed in branchial arches (5-HT2B) could be a fascinating possibility for future therapeutical approaches. In conclusion, studies in *Xenopus* can add a solid background of knowledge that, combined with other approaches, help contributing to the big picture of deciphering the complex roles of serotonin in development.

1. Introduction: An Overview of Serotonin Diverse Functions

Serotonin (5-HT) is a biogenic monoamine that appeared very early in evolution and is present in different phyla, where it is implicated in several physiological and pathological processes. Firstly identified in 1937 as the secretory product of enterocromaffin cells ("enteramine") and later as a serum constricting substance ("serotonin"), it turned out that the two substances were the same molecule, 5-hydroxytryptamine (Whitaker-Azmitia, 1999). But only in recent years 5-HT was appreciated as one of the most evolutionarily fascinating molecules, being implicated in a number of astonishingly diverse functions. 5-HT is a polyfunctional signaling molecule, acting both in developing and mature animals as a neurotransmitter within both the central and peripheral nervous systems, endocrine hormone and growth factor (Barnes and Sharp, 1999; Berger et al., 2009; Gershon, 2013). The 5-HT signaling system is widely conserved, so researches in different animal systems have general relevance for all vertebrates, including humans.

Several organs and cell types can synthesize and/or store 5-HT, examples are the pineal gland, where 5-HT is a precursor of melatonin, the enterochromaffin and myenteric cells in the gut (Côté et al., 2007) and the bone (Tsapakis et al., 2012). These sources of 5-HT are independently regulated and act locally to control the function of the organ to which they belong.

Collectively, tissue sources outside of the brain produce about 95% of 5-HT. On the other hand, virtually all the 5-HT in blood is carried in platelets, which do not make 5-HT, but take it up via the serotonin transporter (SERT): this mechanism makes platelets the fundamental regulators of plasma 5-HT concentration. In the brain, neurons in the raphe nuclei region produce 5-HT (reviewed by Kiyasova and Gaspar, 2011). These neurons project into the cortex and hippocampus and influence an enormous network of excitatory and inhibitory neurotransmission involved in regulating mood, appetite, memory, learning, and other cognitive functions (reviewed by Berger et al., 2009; Deneris and Wyler, 2012).

Tryptophan hydroxylase (Tph) is the rate-limiting enzyme in 5-HT synthesis, which catalyzes the hydroxylation of L-tryptophan. Two isoforms (Tph1 and Tph2) having tryptophan hydroxylating activity were identified. *Tph2* mRNA was reported to be preferentially expressed in the brain. In contrast, *Tph1* mRNA is barely detected in the brain and instead mainly expressed in the pineal gland and in the periphery (Mosienko et al., 2014).

Serotonin diverse effects are mediated by a number of receptors distributed throughout the body. To date, at least fourteen different serotonin receptor subtypes have been identified in mammals and are grouped into seven families (5-HT_1-5-HT_7) (Hannon and Hoyer, 2008). All of the serotonin receptors are G-protein coupled receptors (GPCRs) except the 5-HT_3 ligand-gated ion channel. In the last years, many studies demonstrated that 5-HT receptors can form homo-oligomers, hetero-oligomers and multimers. The formation of these complexes plays important roles in regulating receptors expression, trafficking, clustering, ligand binding, and second messenger activation (Herrick-Davis, 2013).

Serotonin synthesis and uptake become active at early stages during embryogenesis (cf. Gaspar et al., 2003; Smidt and van Hooft, 2013) and the serotonergic system is among the earliest to develop in the CNS (Whitaker-Azmitia, 2001; Homberg et al., 2010). In *Xenopus* embryos 5-HT is present as a maternal pool in the eggs (Fukumoto et al., 2005). In mammals, recent evidence indicates that the placenta is involved in the synthesis of 5-HT from maternally derived tryptophan. It has been suggested that 5-HT synthesized by Tph1 in the murine placenta behaves like a hormone and it is critical in fetal brain development in influencing behavior later in life (Bonnin and Levitt, 2011; Velasquez et al., 2013). It is now clear that in embryonic development 5-HT plays a role as a morphogen-like signal before it acts as a neurotransmitter: it is in fact involved in the establishment of left-right asymmetry (Beyer et al., 2012; Vandenberg et al., 2013) and, through the 2B receptor, in eye, craniofacial and cardiac morphogenesis (Nebigil et al., 2000; Reisoli et al., 2010; Ori et al., 2013).

Paradoxically, less is known on the role of serotonin as a neurodevelopmental factor. Disruption of the 5-HT signaling system is a key developmental component for a number of neuropsychiatric disorders, such as schizophrenia, affective disorders, anxiety, and autism (cf. Kinast et al., 2013; Olivier et al., 2013, Kepser and Homberg, 2014). Nevertheless, the developing serotonergic system and especially its role during brain maturation are still poorly understood.

In the attempt to decipher serotonin role in brain development, a large number of hyposerotonergic genetic models have been generated over the past few years (reviewed by Trowbridge et al., 2011; Lesch et al., 2012; Mosienko et al., 2014). Serotonin depletion has been obtained via targeting of genes involved in 5-HT synthesis (*Tph1* and *Tph2*), specification and determination of the 5-HT phenotype during development (*GATA3*, *Pet1*, and *Lmx1b*), and 5-HT storage or clearance (*Vmat2* and *SERT*). Because the blood brain barrier is not fully functional before postnatal day 12 (Ribatti et al., 2006), 5-HT from the periphery can easily enter the brain and vice versa during the embryonic and early postnatal periods. Thus, alternative sources of 5-HT could compensate for deficiencies in 5-HT levels in other areas. For these reasons these genetic models are not fully conclusive.

2. *Xenopus,* a Non-Mammalian Tetrapod Model System

Xenopus (a genus of African frogs that are fully aquatic, not expensive and easy to maintain in captivity) are robust frogs that tolerate a wide range of living conditions.

As a model system, *Xenopus* embryos combine embryological accessibility with a broad range of functional assays (microsurgery, tissue transplantations, and other manipulations) that are not as easily performed with other vertebrate embryos such as zebrafish, mouse, and rats (reviewed in Wheeler and Brändli 2009; Harland and Grainger, 2011; Abu-Daya et al., 2012; Ori et al., 2013; Schmitt et al., 2014).

Frog eggs are large (~1.2mm diameter), produced in large quantities all over the year and easy to manipulate. Egg production can be stimulated by injection of human chorionic gonadotropin.

Embryos develop externally in simple salt solutions and tolerate extensive manipulation (e.g., single cell, germ layer dissections, tissue transplantations). The developing larvae and tadpoles become transparent facilitating the detection of tissue and organ defects by visual inspection under a dissection microscope and their skin is permeable to small organic molecules. Eggs and embryos are easy to inject with a range of materials (e.g., nucleic acids, proteins, whole nuclei) into whole embryo or specific cells. Cell fate of each early embryonic cell is known, allowing targeted gene knockdown and overexpression studies. Within a couple of days, a tadpole has a fully functional set of organs, and it can be examined to determine if any experimental intervention has had an effect. Finally, *Xenopus* is amenable for large-scale genetic, drug and phenotypic screens that have identified genes involved in diverse developmental and physiological processes and it's a model for rare human inherited diseases (e.g., Parain et al., 2011; Zhang et al., 2013; Schmitt et al., 2014).

Two species are mostly used, *Xenopus laevis*, the species first widely used by researchers, and *Xenopus tropicalis*, more recently adopted because of advantages for genetic and genomic work. In contrast with *tropicalis,* which is a true diploid (Hirsch et al., 2002), the genome of *X. laevis* is paleotetraploid. However, a subsequent diploidization process has returned the genome of *X. laevis* to a genetically diploid state (Uno et al., 2013). *X. laevis* remains the preferred *Xenopus* species for many experiments because experimental manipulation and microsurgery is easier due to larger (about five times) oocytes, eggs, and embryos. Finally, *X. laevis* embryos develop at room temperature, which simplifies experimental handling and husbandry.

Public databases such as Xenbase harbor a vaste array of tools and data useful for researchers, containing information about all aspects of research involving the two primary *Xenopus* species (Klein et al., 2002; James-Zorn et al., 2013). The genome of *X. tropicalis* has been completely sequenced. *Xenopus* is a tetrapod in the same evolutionary branch of mammals and its genome displays remarkable structural similarity with the human genome (Hellsten et al., 2010) meaning that findings from *Xenopus* provide insights into many human conditions and diseases.

One of the advantages of the *Xenopus* model is the possibility to perform gene gain and loss of function experiments in an easy, effective and extremely rapid fashion. Embryos cleavage pattern allows injecting the mRNA in just one side of the embryo (Figure 1). For this unique feature the injected side can be immediately compared with the control side in the same embryo. A detailed fate map of each blastomere has been described allowing to direct the microinjection in just the presumptive territory of interest, thus minimizing the possible side effects of a widespread mRNA overexpression that represents a possible limitation in other models.

The gene loss-of-function strategy based on injection of antisense morpholino oligonucleotides (MO) is at the moment the most popular way to disrupt a gene function in *Xenopus* embryos, and has been successfully employed in several laboratories including ours (e.g., Reisoli et al., 2008; Reisoli et al., 2010; Ori et al., 2013) (Figure 1).

MOs are nucleic acids characterized by morpholine rings, which replace the ribose or deoxyribose moieties. The mechanism of action of MOs relies on the stability of the duplex formed on the target RNA that leads to steric hindrance. A second strategy makes use of splicing morpholinos. By blocking sites involved in pre-mRNA splicing, morpholinos can be used to modify and control normal splicing events. This activity, producing aberrant or truncated proteins, can be conveniently assayed by RT-PCR, with successful splice-modification appearing as changes in the RT-PCR product band.

The absence of off-target effects can be assessed by performing functional rescue experiments (see for example Reisoli et al., 2010). MO-mediated translational inhibition has become the antisense oligonucleotide chemistry of choice for gene knockdowns in non-mammalian embryos due to the efficiency and specificity, resistance to nucleases, good solubility in aqueous solutions, low toxicity, and low production costs (c.f. Eisen and Smith, 2008).

Recently, a number of powerful genome editing tools have emerged, including zinc-finger nucleases (ZFNs), transcription activator-like effector nucleases (TALENs), and the RNA guided CRISPR/Cas nuclease system, that enable the precise engineering of diverse model systems and appears promising in producing mutants in systems refractory to genetics.

All three genome editing technologies have been successfully adapted to *Xenopus* embryos (cf. Schmitt et al., 2014) and several reports document the effectiveness of the CRISPR/Cas system at modifying the *X. tropicalis* genome and the successful germline transmission of the CRISPR/Cas-induced mutations to the F1 generation (Bliz et al., 2013; Nakayama et al., 2013; Guo et al., 2014).

Also, recent advances in sequencing technology such as the RNA-seq have made possible genome-wide analyses in organisms without a published genome sequence. This new and potent approach has been recently demonstrated to be suitable for analyzing the *Xenopus leavis* transcriptome (Blower et al., 2013; Amin et al., 2014).

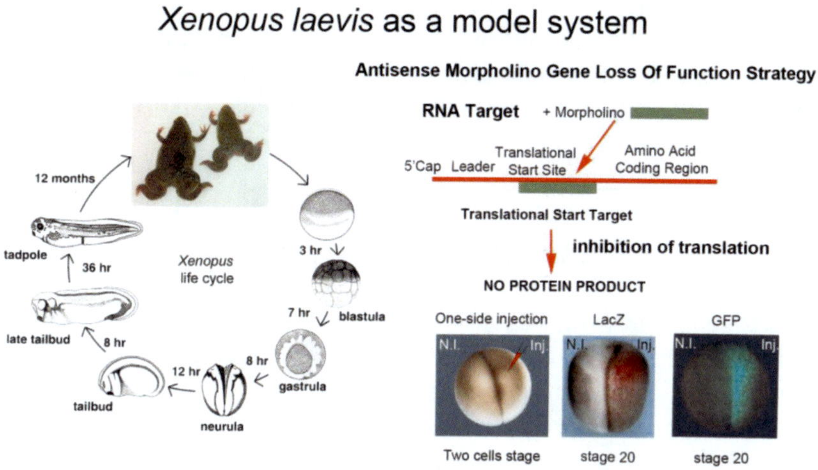

Reproduced with permission from Ori et al., The International Journal of Developmental Biology (Int. J. Dev. Biol.) (2013) 57: 707-714.

Figure 1. *Xenopus laevis* as a model system to study gene function. On the left, the life cycle of *Xenopus laevis.* On the right, the gene loss of function strategy that specifically blocks the translation of a gene of interest by microinjecting specific morpholino oligos in one side of the embryo. Antisense morpholino is targeted to sequences close to the translation start site. The injected side is visualized by the co-injection of a reporter mRNA such as LacZ or GFP. Examples of chromogenic reaction for beta galactosidase (red) and GFP (green fluorescence) are shown in neural stage embryos.

3. Serotonin Developmental Functions: What We Learned from *Xenopus*

The presence of 5-HT, its receptors and the serotonin transporter during development as well as the ability of 5-HT-specific pharmacological agents to interfere with embryonic development suggested that 5-HT could function as a humoral morphogen before being a neurotransmitter. Since the pioneer studies of Buznikov and his collaborators (reviewed in Buznikov, 2007), this hypothesis has been pursued by several laboratories (see for example Choi et al., 1997; Azmitia, 2001; Buznikov et al., 2001; Gaspar et al., 2003; De Lucchini et al., 2005). Understanding the role of 5-HT in development could also be critical to identifying the possible effects of SSRI on the human fetus, since these are the drugs of choice for the treatment of depression as well as for a wide spectrum of other mood and behavioural disorders, in pregnant and lactating women (see for example Hanley and Oberlander, 2014). SSRI may pass to the fetus through the placenta and neonate through breastfeeding, potentially exposing them to increased level of serotonin. Here we report some recent advances on the developmental role of 5-HT, obtained thanks to studies in *Xenopus* embryos.

Establishment of Left/Right Asymmetry

The establishment of laterality is one of the most fascinating fields in developmental biology and there is now strong evidence that asymmetry is an ancient, highly conserved feature of single cells and higher organisms. Defects in left-right (LR) asymmetry are among the congenital birth malformations and although they affect less live births compared with other defects, they have severe medical consequences. For example, in addition to internal organs defects, altered neurological asymmetries have been linked to diseases and disabilities such as dyslexia and schizophrenia (e.g., Corballis, 2012).

Asymmetric organ morphogenesis is initiated during embryogenesis. In early vertebrate neurula embryos, a gene called *Nodal* (*Xnr-1* in *Xenopus*) becomes asymmetrically expressed and starts the so-called Nodal cascade which is a conserved feature of all vertebrates (Yoshiba and Hamada, 2014). But how this cascade is initiated is still not deciphered.

So far two models to explain the establishment of LR asymmetry have been proposed. The first one, the so called "flow model", suggests that symmetry is first broken during neurulation when cilia, localized in the gastrocoel roof plate (the node homologous structure in *Xenopus*), generate a chiral fluid flow that asymmetrically distributes a morphogen or causes asymmetric bending of sensory cilia. The second one, the "ion-flux model", proposes a much earlier phenomenon that drives asymmetric localization of ion transporters thus establishing biased voltage gradient resulting in asymmetric localization of a charged molecule required for left-right identity (Vanderberg et al., 2012; Blum et al., 2014).

The amphibian model system has historically been used to study left-right patterning since the early 1900s by Spemann and colleagues and a function for serotonin signaling in LR axis specification in *Xenopus laevis* has been inferred several years ago from a drug screen, in which embryos treated with serotonin receptors antagonists developed LR defects (Fukumoto et al., 2005). The precise molecular mechanisms of symmetry breaking is still under debate and far beyond the scope of this review, as several excellent reviews have been published (as for example Vandenberg et al., 2013a,c; Blum et al., 2014) but what is worth to note is that the two main models may converge on serotonin as key molecule involved in this process.

Beyer and colleagues (2012), in fact, suggested a permissive role for serotonin in the specification of the gastrocoel roof plate and cilia dependent fluid flow while Vandenberg and colleagues (2013b) reported that the charged molecule serotonin is redistributed asymmetrically along the left-right axis at cleavage stages to provide, in the very early steps of development, the identity to blastomeres.

Given that in all models of symmetry breaking the serotonin appears to be a central actor, understanding the precise timing of symmetry breaking, i.e., when serotonin actually acts during LR patterning, is crucial also because it identifies the embryonic stages that would be most sensitive to the serotonergic compounds that are in widespread medical use (Homberg et al., 2010).

The frog embryo has been proven to be a powerful model in which to study the earliest events that lead to asymmetric gene expression revealing again the versatile nature of serotonin in regulating so many different developmental steps even in pre-neural stages.

Developmental Role of 5-HT2B Receptor

In *Xenopus* embryos 5-HT is present as a maternal pool in the eggs and the main components of the serotonergic system are expressed during the earliest stages of embryonic development (Fukumoto et al., 2005; Nikishin et al., 2012). In the CNS 5-HT is first produced by serotonergic neurons of the raphe nuclei at tailbud stages (van Mier et al., 1986). Serotonergic cells are found in several other tissues or organs of the *Xenopus* larvae, such as the retina, the skin and the gut (Ori et al., 2013), (Figure 2).

5-HT2B is the only 5-HT receptor that has clearly been shown to have an early developmental role in mice, where it is expressed in proliferating heart cells and in neural crest-derived progenitors (Choi et al., 1997; Nebigil et al., 2000). The targeted inactivation of the 5-HT2B gene leads to embryonic and neonatal death caused by heart defects.

These studies, also supported by in vitro analyses, demonstrated that 5-HT via 5-HT2B receptor is an important regulator of cardiac myocytes differentiation, proliferation and survival (Nebigil et al., 2000, 2003).

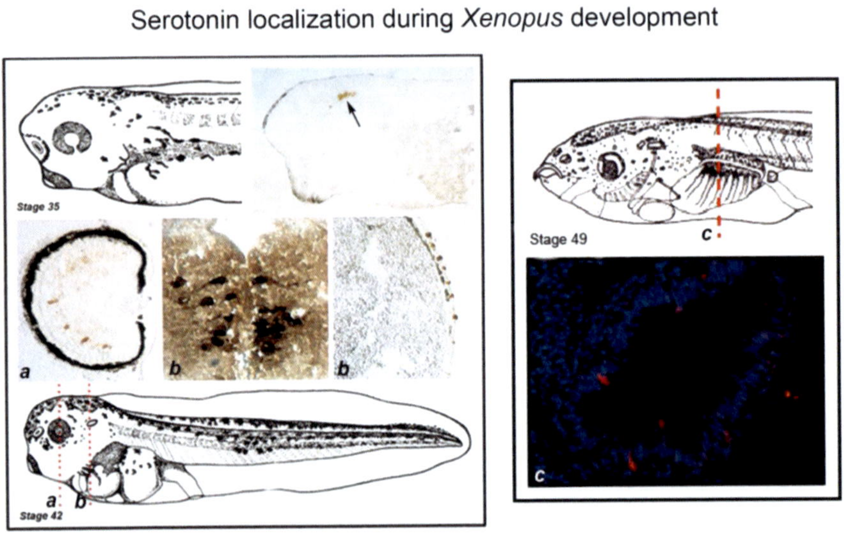

Reproduced with permission from Ori et al., The International Journal of Developmental Biology (Int. J. Dev. Biol.) (2013) 57: 707-714.

Figure 2. Visualization of serotonergic cells during *Xenopus* embryogenesis. Distribution of serotonergic cells in *Xenopus* larvae, as detected in cryostat sections by immunostaining with an anti-serotonin antibody. Left panel, upper row: sagittal section of a late tailbud (stage 35) larva showing serotonin cells in the raphe nuclei (arrow); middle row, from left to right: serotonergic cells are found in the retina (amacrine cells), in the raphe nuclei, and in the skin of a late tailbud (stage 42) larva. Right panel: coronal section showing serotonin cells detected in the gut of a stage 49 tadpole (red fluorescence). Coronal planes are indicated as a, b, c.

In *Xenopus*, pharmacological treatments with 5-HT2B antagonists resulted in reproducible phenotypic defects including reduced pigmentation compared to the controls, heart and gut defects with reduced heartbeats and edema in the pericardial region. Heart defects closely resemble what found in mice. Consistently with what was observed in the 5-HT2B knockout mouse, *Xenopus* 5-HT2B morphants exhibited extensive apoptosis in the heart at larval stages (De Lucchini et al., 2005). Interestingly, increased risk for congenital heart defects after exposure to SSRI has been reported in numerous studies (Hanley and Oberlander, 2014).

Activation of 5-HT2B receptors is a limiting step in the development of pulmonary hypertension. In both humans and mice, pulmonary hypertension (a progressive and often fatal disorder in humans that results from an increase in pulmonary blood pressure associated with abnormal vascular proliferation) is associated with a substantial increase in 5-HT2B receptor expression in pulmonary arteries (Launay et al., 2002). Again, several cohort studies have reported a relationship between maternal antidepressant use during pregnancy and neonatal persistent pulmonary hypertension (PPHN), although further research is needed to clarify the relationship between prenatal SSRI exposure and PPHN (Hanley and Oberlander, 2014).

Developmental expression of the 5-HT2B receptor in *Xenopus* is associated with proliferative regions of the developing brain and retina, and includes the branchial arches of tailbud embryos and the periocular mesenchyme (De Lucchini et al., 2003; Reisoli et al., 2008, 2010).

Although a morphogenetic role for the 5-HT2B receptor was suspected because of the potent teratogenic activity of its antagonists in cultured mouse embryos, a direct demonstration of its morphogenetic role was obtained from studies in *Xenopus*, as we review in the following sections.

5-HT2B Receptor in Eye Morphogenesis

In vertebrates, eye development comprises inductive and morphogenetic events that are finely regulated by the coordinated action of many intrinsic and extrinsic factors. Recent evidence in *Xenopus* demonstrates that serotonin signaling can influence eye development (De Lucchini et al., 2005; Reisoli et al., 2008; Ori et al., 2013). Serotonin, acting via the 5-HT2B receptor, is involved in the control of retinoblasts proliferation and survival in *Xenopus* retinogenesis and it is required for the correct formation of retinal cytoarchitecture and layering. The in vivo experiments revealed that dysregulation of serotonin signaling, via 5-HT2B receptors, resulted in retinal dysplasia, microftalmia, coloboma and the formation of eyes with an irregular form, position and orientation (Figure 3b-e). The same studies revealed for the first time a specific expression of the 5-HT2B receptor mRNA in the periocular mesenchyme (POM), that represents a key signaling center required for a correct eye morphogenesis (Reisoli et al., 2008; Ori et al., 2013). POM is known to provide multiple cell lineages necessary for normal ocular development as well as essential signals for the patterning of ocular primordia including the morphogenetic extension of the optic stalk, the anterior eye segment development and the optic fissure closure (Gage et al., 2005). Dysregulation of 5-HT2B gene expression, due to both gene gain or loss of function, produces defects in the morphology, position and orientation of the eye in the larval head that cannot be directly explained by mechanisms involved in retinal histogenesis, but are consistent with a role of 5-HT as one of the morphogens regulating POM functions. The finding that expression of the 5-HT2B receptor is associated with proliferative regions not only in the retina but also in the developing brain is suggestive of a possible involvement of 5-HT signaling in brain development (De Lucchini et al., 2003).

Shaping the First Branchial Arch: Requirement of Serotonin Signaling in Craniofacial Morphogenesis and Jaw Joint Formation

Craniofacial morphogenesis is a complex process and the blueprint for building the vertebrate head and face is established very early during embryonic development. In fact in humans, the most critical steps take place between about the third to eighth weeks of pregnancy. The facial region of the embryo starts out as a relatively blank slate, consisting of three distinct tissue layers known as ectoderm (outside), mesoderm (middle), and endoderm (inside) lying adjacent to one another. Shortly thereafter, the most critical step of head and facial development occurs: the formation of neural crest cells (NCC) (Trainor and Krumlauf, 2000).

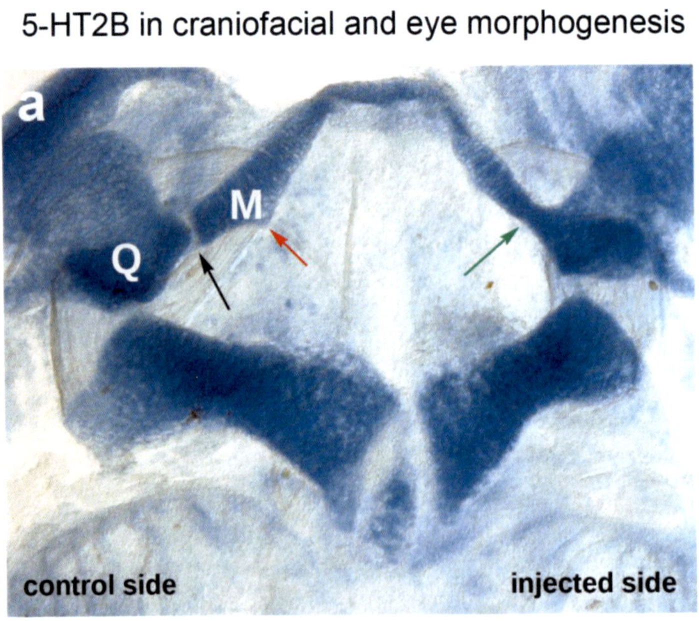

Figure 3. (Continued).

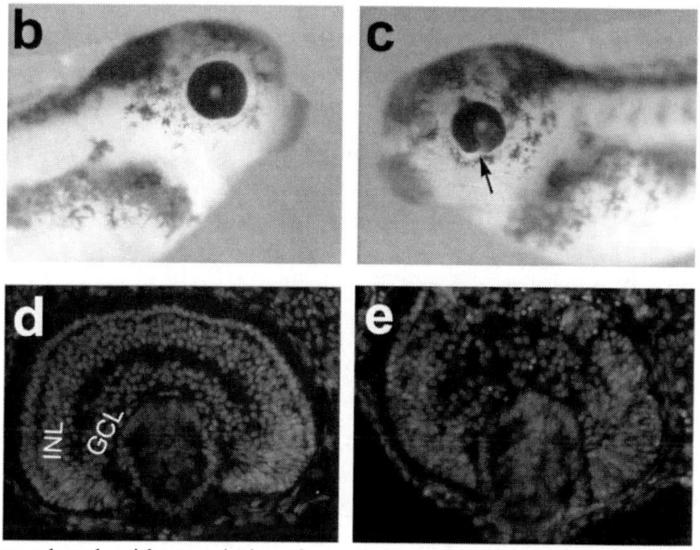

(b-e) Reproduced with permission from Ori et al., The International Journal of Developmental Biology (Int. J. Dev. Biol.) (2013) 57: 707-714.

Figure 3. 5-HT2B in craniofacial and eye morphogenesis. (a) Role of serotonergic signaling in the first branchial arch development. Craniofacial skeletal alterations in 5-HT2B morphants (st. 49). Alcian blue staining of a flat mount embryo showing the jaw joint region of the control and injected side. The jaw joint (black arrow in the control side) is lacking in the injected side (green arrow); note also the reduction of the quadrate (Q) and the absence of the ventral cartilaginous muscular process of the Meckel's cartilage, which is present in the control side (red arrow). Q, quadrate; M, Meckel's cartilage. (b-e) Role of serotonergic signaling during eye development. Effects of 5-HT2B morpholino injection on *Xenopus* eye development. Lateral (b,c) views of control and injected side of the same tadpole (st. 42). The arrow in (c) points to the optic fissure that fails to close ventrally. (d,e) Hoechst nuclear staining (blue) of cryostat eye cross-sections of control and morpholino-injected stage 42 tadpole: note the altered retinal layering in (e). INL, inner nuclear layer; GCL, ganglion cell layer.

Cranial NCC are derived from the neural ectoderm and migrate within branchial arches where they finally differentiate originating cartilage, bone, connective tissue, sensory neurons, glia, and pigments cells.

Signals from the endoderm and the ectoderm, which overlie the branchial arches, are crucial for correct craniofacial development and in particular for the specification of the first branchial arch (Minoux and Rijli, 2010).

In the last years the serotonin receptor 2B, 5-HT2B, was enumerated among the complex interactive network of extrinsic factors that regulates craniofacial skeleton development. 5-HT2B was revealed to be required for the correct first branchial arch derivatives development in *Xenopus* (Reisoli et al., 2010; Ori et al., 2013). *5-HT2B* mRNA is present in the branchial arches of late tailbud *Xenopus* embryos, as well as in the branchial arches of the mouse embryo at ~E9 (Lauder et al., 2000; Reisoli et al., 2010).

However, developmental craniofacial defects in 5-HT2B knockout mouse embryos have not been reported, probably because the most penetrating phenotype of 5-HT2B mutant mice leads to early embryonic death in uterus (Nebigil et al., 2000). Knockdown of 5-HT2B in *Xenopus* embryos resulted in *Xbap* (also known as *Nkx3.2*) downregulation, a gene codifying for a transcription factor expressed in the precursor cells of the jaw joint region, and in a fusion of a hypomorphic quadrate with the Meckel's cartilage into a single element, leading to the loss of the jaw joint (Figure 3a). The 5-HT2B signaling is therefore required to define and sustain the *Xbap* gene expression necessary for the jaw and jaw joint formation (Reisoli et al., 2010). Another signaling pathway known to be able to positively control *Bapx1* expression is endothelin 1 (Edn1) (Walker et al., 2007).

Since both endothelin and 5-HT2B signaling act through the PLC beta 3 effector, they could cooperate in modulating *Xbap* expression and reinforcing the PLCB3 pathway in cranial NCC during the first arch patterning (Reisoli et al., 2010). *PLCB3* is the orthologue of the mammalian *PLCB4* gene, which is implicated in the Auriculo-condylar Syndrome (ACS), a birth defect involving in particular the first branchial arch derivatives (Rieder et al., 2012). Missense mutations located in the catalytic domain of PLCB4 were identified in patients affected by the syndrome. These mutations could negatively modulate the endothelin pathway and, in turn, down regulate key genes such as *Dlx5* and *Dlx6*. Interestingly, Reisoli and colleagues demonstrated that the endothelin receptor and 5-HT2B signaling might converge on the PLCB3 but they act in a non-redundant manner and that 5-HT2B can regulate *Dlx6* gene expression in the branchial arches (Reisoli et al., 2010; Ori et al., unpublished results).

It is tempting to imagine that 5-HT2B could be a new and important pharmacological target for future therapies for craniofacial congenital defects of the first branchial arch, as several 5-HT2B agonists are already commercially available.

The discovery of serotonin involvement in shaping the first branchial arch represents a landmark in vertebrate evolution, as the emerging of distinct elements of the jaws articulated by a jaw joint allowed the acquisition of a buccal skeleton that was the major evolutionary novelty of vertebrates that shifted from a lifestyle of passive filter feeding to one of active predation (Gans and Northcutt, 1983).

Conclusion

Here we learned that the frog model allowed substantial advances in unveiling unexpected and important roles of serotonin during development relevant for both evolutionary and medical perspectives.

Given its implications in several mental illnesses, serotonin is one of the most studied human neurotransmitters. Up to 20% of pregnant women suffer from depression and pharmacotherapy is often necessary.

Consequently, recent attention has focused on the possible risks of antidepressant treatment during pregnancy. SSRI, which are among the mostly prescribed antidepressant drugs, inhibit the function of the serotonin transporter (5-HTT) leading to an accumulation of 5-HT in the extracellular space. This increases the magnitude and duration of the activity of 5-HT on pre- and postsynaptic 5-HT receptors. SSRI cross freely the placental barrier and are thus transferred to the fetus as well as to the newborn during lactation. Thus a significant number of unborn and newborn children may be exposed to increased level of serotonin during critical phases of development and this maternal-fetal interactions may have long-term health implications (Kepser and Homberg, 2014; Hanley and Oberlander, 2014).

Evidence of the last few years show that serotonin signaling can influence many developmental processes both in the central nervous system and in the peripheral organs (De Lucchini et al., 2005; Reisoli et al., 2010; Migliarini et al., 2013, Mosienko et al., 2014). In mammals the possible role of 5-HT in the early stage of brain development is not definitively demonstrated as it is eclipsed by serotonin produced from external sources, such as the placenta, that can reach the brain through the immature blood brain barrier. In contrast, the total level of 5-HT in fertilized *Xenopus* eggs (3.11 pmol/embryo) rapidly decreases to less than 0.16 pmol/embryo at gastrula stage, while it raises again from late tailbud stages, because of zygotic serotonin synthesized by the developing nervous system (Fukumoto et al., 2005). We can therefore imagine that *Xenopus* could be conveniently used as a tetrapod model for a comprehensive picture of serotonin function during CNS development without the influence of the maternal 5-HT. This kind of analysis is not easily done in mammals, where the contribution of maternal serotonin cannot be completely avoided. We can envisage that this approach, reinforced by the use of recent technologies such as high throughput sequencing and novel genome editing tools, could open multiple new routes of investigations and further strong advances in the field, where *Xenopus* can be conveniently employed as first-line animal model.

References

Abu-Daya, A., Khokha, M. K., Zimmerman, L. B. (2012). The Hitchhiker's Guide to *Xenopus* Genetics. *Genesis* 50, 164-175.

Amin, N. M., Tandon, P., Osborne Nishimura, E., Conlon, F. L. (2014). RNA-seq in the tetraploid Xenopus laevis enables genome-wide insight in a classic developmental biology model organism. *Methods* 66, 398-409.

Azmitia, E. C. (2001). Modern views on an ancient chemical: serotonin effects on cell proliferation, maturation, and apoptosis. *Brain Res. Bull.* 56, 413-24.

Barnes, N. M., Sharp, T. (1999). A review of central 5-HT receptors and their function. *Neuropharmacology* 38, 1083-1152.

Berger, M., Gray, J. A., Roth, B. L. (2009). The expanded biology of serotonin. *Annu. Rev. Med.* 60: 355-366.

Beyer, T., Danilchick, M., Thumberger, T., Vick, P., Tisler, M., Schneider, I., Bogusch, S., Andre, P., Ulmer, B., Walentek, P., Niesler, B., Blum, M., Schweickert, A. (2012). Serotonin signaling is required for Wnt-dependent GRP specification and leftward flow in Xenopus. *Curr. Biol.* 22: 33-39.

Blitz, I. L., Biesinger, J., Xie, X., Cho, K. W. (2013). Biallelic genome modification in F(0) Xenopus tropicalis embryos using the CRISPR/Cas system. *Genesis* 51, 827-34.

Blower, M. D., Jambhekar, A., Schwarz, D. S., Toombs, J. A. (2013). Combining different mRNA capture methods to analyze the transcriptome: analysis of the Xenopus laevis transcriptome. *PLoS One* 8 (10):e77700 eCollection 2013.

Blum, M., Feistel, K., Thumberger, T., Schweickert, A. The evolution and conservation of left-right patterning mechanisms. (2014) *Development* 141, 1603-1613.

Bonnin, A., Goeden, N., Chen, K., Wilson, M. L., King, J., Shih, J. C., Blakely, R. D., Deneris, E. S., Levitt, P. (2011). A transient placental source of serotonin for the fetal forebrain. *Nature* 472: 347-350.

Buznikov, G. A., Lambert, H. W., Lauder, J. M. (2001). Serotonin and serotonin-like substances as regulators of early embryogenesis and morphogenesis. *Cell Tissue Res.* 305, 177-186.

Buznikov, G. A. (2007). Preneural transmitters as regulators of embryogenesis. Current state of the problem. *Russ. J. Dev. Biol.* 38: 213-220.

Choi, D. S., Ward, S. J., Messaddeq, N., Launay, J. M., Maroteaux, L. (1997). 5-HT2B receptor-mediated serotonin morphogenetic functions in mouse cranial neural crest and myocardiac cells. *Development* 124, 1745-1755.

Corballis, M. C. (2012). Lateralization of the human brain. *Prog. Brain Res.* 195, 103-21.

Côté, F., Fligny, C., Bayard, E., Launay, J. M., Gershon, M. D., Mallet, J. Vodjdani, G. (2007). Maternal serotonin is crucial for murine embryonic development. *Proc. Natl. Acad. Sci. US* 104, 329-334.

De Lucchini, S., Ori, M., Nardini, M., Marracci, S., Nardi, I. (2003). Expression of 5-HT2B and 5-HT2C receptor genes is associated with proliferative regions of Xenopus developing brain and eye. *Brain Res. Mol. Brain Res.* 115, 196-201.

De Lucchini, S., Ori, M., Cremisi, F., Nardini, M., Nardi, I. (2005). 5-HT2B-mediated serotonin signaling is required for eye morphogenesis in Xenopus. *Mol. Cell. Neurosci.* 29, 299-312.

Deneris, E. S., Wyler, S. C. (2012). Serotonergic transcriptional networks and potential importance to mental health. *Nat. Neurosci.* 15, 519-27.

Eisen, J. S., Smith, J. C. (2008) Controlling morpholino experiments: don't stop making antisense, *Development* 135, 1735-43.

Fukumoto, T., Kema, I. P., Levin, M. (2005). Serotonin signaling is a very early step in patterning of the left-right axis in chick and frog embryos. *Curr. Biol.* 15, 794-803.

Gage, P. J., Rhoades, W., Prucka, S. K., Hjalt, T. (2005). Fate maps of neural crest and mesoderm in the mammalian eye. *Invest. Ophthalmol. Vis. Sci.* 46, 4200-4208.

Gans, C., Northcutt, R. G. (1983) Neural crest and the origin of vertebrates: a new head. *Science* 220, 268-73.

Gaspar, P., Cases, O., Maroteaux, L. (2003). The developmental role of serotonin: news from mouse molecular genetics. *Nat. Rev. Neurosci.* 4, 1002-1012.

Gershon, M. D. (2013). 5-Hydroxytryptamine (serotonin) in the gastrointestinal tract. *Curr. Opin. Endocrinol. Diabetes Obes.* 20, 14-21.

Guo, X., Zhang, T., Hu, Z., Zhang, Y., Shi, Z., Wang, Q., Cui, Y., Wang, F., Zhao, H., Chen, Y. (2014). Efficient RNA/Cas9-mediated genome editing in Xenopus tropicalis. *Development* 141, 707-14.

Hanley, G. E., Oberlander, T. F. (2014). The effect of perinatal exposures on the infant: Antidepressants and depression. *Best Practice and Research Clinical Obstetrics and Gynaecology* 28 37-48.

Hannon, J., Hoyer, D. (2008). Molecular biology of 5-HT receptors. *Behav. Brain Res.* 195, 198-213.

Harland, R. M., Grainger, R. M. (2011) Xenopus research: metamorphosed by genetics and genomics. *Trends Genet.* 27, 507-15.

Hellsten, U., Harland, R. M., Gilchrist, M. J., Hendrix, D., Jurka, J., Kapitonov, V., Ovcharenko, I., Putnam, N. H., Shu, S., Taher, L., Blitz, I. L., Blumberg, B., Dichmann, D. S., Dubchak, I., Amaya, E., Detter, J. C., Fletcher, R., Gerhard, D. S., Goodstein, D., Graves, T., Grigoriev, I. V., Grimwood, J., Kawashima, T., Lindquist, E., Lucas, S. M., Mead, P. E., Mitros, T., Ogino, H., Ohta, Y., Poliakov, A. V., Pollet, N., Robert, J., Salamov, A., Sater, A. K., Schmutz, J., Terry, A., Vize, P. D., Warren, W. C., Wells, D., Wills, A., Wilson, R. K., Zimmerman, L. B., Zorn, A. M., Grainger, R., Grammer, T., Khokha, M. K., Richardson, P. M., Rokhsar, D. S. (2010). The genome of the Western clawed frog Xenopus tropicalis. *Science* 328, 633-6.

Herrick-Davis, K. (2013). Functional significance of serotonin receptor dimerization. *Experimental Brain Research* 230, 375-386.

Hirsch, N., Zimmerman, L. B., Grainger, R. M. (2002) Xenopus, the next generation: X. tropicalis genetics and genomics, *Dev. Dyn.* 225 422-433.

Homberg, J. R., Schubert, D., Gaspar, P. (2010). New perspectives on the neurodevelopmental effects of SSRIs. *Trends Pharmacol. Sci.* 31, 60-65.

James-Zorn, C., Ponferrada, V. G., Jarabek, C. J., Burns, K. A., Segerdell, E. J., Lee, J., Snyder, K., Bhattacharyya, B., Karpinka, J. B., Fortriede, J., Bowes, J. B., Zorn, A. M., Vize, P. D. (2013) Xenbase: expansion and updates of the Xenopus model organism database. *Nucleic Acids Res.* 41 D865-D870.

Kepser, L. J., Homberg, J. R. (2014). The neurodevelopmental effects of serotonin: A behavioural perspective. *Behav. Brain Res. doi*: 10.1016/j. bbr.2014.05.022. [Epub. ahead of print].

Kinast, K., Peeters, D., Kolk, S. M., Schubert, D., Homberg, J. R. (2013). Genetic and pharmacological manipulations of the serotonergic system in early life: neurodevelopmental underpinnings of autism-related behavior. *Front. Cell. Neurosci.* 7:72.

Kiyasova, V., Gaspar, P. (2011). Development of raphe serotonin neurons from specification to guidance. *Eur. J. Neurosci.* 34, 1553-62.

Klein, S. L., Strausberg, R. L., Wagner, L., Pontius, J., Clifton, S. W., Richardson, P. (2002). Genetic and genomic tools for Xenopus research: the NIH Xenopus initiative, *Dev. Dyn.* 225, 384-391.

Lauder, J. M., Wilkie, M. B., Wu, C., Singh, S. (2000). Expression of 5-HT(2A), 5-HT(2B) and 5-HT(2C) receptors in the mouse embryo. *Int. J. Dev. Neurosci.* 18, 653-662.

Launay, J. M., Hervé, P., Peoc'h, K., Tournois, C., Callebert, J., Nebigil, C. G., Etienne, N., Drouet, L., Humbert, M., Simonneau, G., Maroteaux, L. (2002). Function of the serotonin 5-hydroxytryptamine 2B receptor in pulmonary hypertension. *Nat. Med.* 8, 1129-35.

Lesch, K. P., Araragi, N., Waider, J., van den Hove, D., Gutknecht, L. (2012). Targeting brain serotonin synthesis: insights into neurodevelopmental disorders with long-term outcomes related to negative emotionality, aggression and antisocial behaviour. *Philos. Trans. R Soc. Lond. B Biol. Sci.* 367, 2426-43.

Migliarini, S., Pacini, G., Pelosi, B., Lunardi, G., Pasqualetti, M. (2013). Lack of brain serotonin affects postnatal development and serotonergic neuronal circuitry formation. *Mol. Psychiatry* 18, 1106-1118.

Minoux, M., Rijli, F. M. (2010). Molecular mechanisms of cranial neural crest cell migration and patterning in craniofacial development. *Development* 137: 2605-21.

Mosienko, V., Beis, D., Pasqualetti, M., Waider, J., Matthes, S., Qadri, F., Bader, M., Alenina, N. (2014) Life without brain serotonin: Reevaluation of serotonin function with mice deficient in brain serotonin synthesis. Behav. *Brain Res.* doi: 10.1016/j.bbr.2014.06.005. [Epub. ahead of print].

Nakayama, T., Fish, M. B., Fisher, M., Oomen-Hajagos, J., Thomsen, G. H., Grainger, R. M. (2013) Simple and efficient CRISPR/Cas9-mediated targeted mutagenesis in Xenopus tropicalis. *Genesis* 51, 835-43.

Nebigil, C. G., Choi, D. S., Dierich, A., Hickel, P., Le Meur, M., Messaddeq, N., Launay, J. M., Maroteaux, L. (2000). Serotonin 2B receptor is required for heart development. *Proc. Natl. Acad. Sci. US* 97, 9508-9513.

Nebigil, C. G., Etienne, N., Messaddeq, N., Maroteaux, L. (2003). Serotonin is a novel survival factor of cardiomyocytes: mitochondria as a target of 5-HT2B receptor signaling. *FASEB J.* 17, 1373-1375.

Nikishin, D. A., Kremnyov, S. V., Konduktorova, V. V., Shmukler, Y. B. (2012). Expression of serotonergic system components during early *Xenopus* embryogenesis. *Int. J. Dev. Biol.* 56, 385-91.

Olivier, J. D., Akerud, H., Kaihola, H., Pawluski, J. L., Skalkidou, A., Högberg, U., Sundström-Poromaa, I. (2013). The effects of maternal depression and maternal selective serotonin reuptake inhibitor exposure on offspring. *Front. Cell. Neurosci.* 7:73.

Ori, M., De Lucchini, S., Marras, G., Nardi, I. (2013). Unraveling new roles for serotonin receptor 2B in development: key findings from Xenopus. *Int. J. Dev. Biol.* 57: 707-714.

Parain, K., Mazurier, N., Bronchain, O., Borday, C., Cabochette, P., Chesneau, A., Colozza, G., El Yakoubi, W., Hamdache, J., Locker, M., Gilchrist, M. J., Pollet, N., Perron, M. (2012). A large scale screen for neural stem cell markers in Xenopus retina. *Dev. Neurobiol.* 72, 491-506.

Reisoli, E., De Lucchini, S., Anelli, T., Biagioni, S., Nardi, I., Ori, M. (2008). Overexpression of 5-HT2B receptor results in retinal dysplasia and defective ocular morphogenesis in Xenopus embryos. *Brain Res.* 1244, 32-9.

Reisoli, E., De Lucchini, S., Nardi, I., Ori, M. (2010). Serotonin 2B receptor signaling is required for craniofacial morphogenesis and jaw joint formation in Xenopus. *Development* 137, 2927-2937.

Ribatti, D., Nico, B., Crivellato, E., Artico, M. (2006). Development of the blood-brain barrier: a historical point of view. *Anat. Rec. B New Anat.* 289, 3-8.

Rieder, M. J., Green, G. E., Park, S. S., Stamper, B. D., Gordon, C. T., Johnson, J. M., Cunniff, C. M., Smith, J. D., Emery, S. B., Lyonnet, S., Amiel, J., Holder, M., Heggie, A. A., Bamshad, M. J., Nickerson, D. A., Cox, T. C., Hing, A. V., Horst, J. A., Cunningham, M. L. (2012). A human homeotic transformation resulting from mutations in PLCB4 and GNAI3 causes auriculocondylar syndrome. *Am. J. Hum. Genet.* 90, 907-14.

Schmitt, S. M., Gull, M., Brändli, A. W. (2014). Engineering Xenopus embryos for phenotypic drug discovery *Adv. Drug Deliv. Rev.* 69-70, 225-46.

Smidt, M. P., van Hooft, J. A. (2013). Subset specification of central serotonergic neurons. *Front. Cell. Neurosci.* 7:200.

Trainor, P., Krumlauf, R. (2000). Patterning the cranial neural crest: hindbrain segmentation and Hox gene plasticity. *Nat. Rev. Neurosci.* 1, 116-124.

Trowbridge, S., Narboux-Neme, N., Gaspar, P. (2011). Genetic models of serotonin (5-HT) depletion: what do they tell us about the developmental role of 5-HT? *The Anatomical Record* 294:1615-1623.

Tsapakis, E. M., Gamie, Z., Tran, G. T., Adshead, S., Lampard, A., Mantalaris, A., Tsiridis, E. (2012). The adverse skeletal effects of selective serotonin reuptake inhibitors. *Eur. Psychiatry* 27: 156-169.

Uno, Y., Nishida, C., Takagi, C., Ueno, N., Matsuda, Y. (2013) Homoeologous chromosomes of Xenopus laevis are highly conserved after whole-genome duplication, *Heredity* 111, 430-436.

Vandenberg, L. N., Levin, M. (2012). Polarity proteins are required for left-right axis orientation and twin-twin instruction. *Genesis* 50, 219-234.

Vandenberg, L. N., Levin, M. (2013a) A unified model for left-right asymmetry? Comparison and synthesis of molecular models of embryonic laterality. *Dev. Biol.* 379, 1-15.

Vandenberg, L. N., Lemire, J. M., Levin, M. (2013b). Serotonin has early, cilia-independent roles in Xenopus left-right patterning. *Dis. Model. Mech.* 6, 261-268.

Vandenberg, L. N., Lemire, J. M., Levin, M. (2013c). It's never too early to get it Right: A conserved role for the cytoskeleton in left-right asymmetry. *Commun. Integr. Biol.* 6(6):e27155.

Van Mier, P., Joosten, H. W., Van Rheden, R., Ten Donkelaar, H. J. (1986). The development of serotonergic raphespinal projections in *Xenopus laevis. Int. J. Dev. Neurosci.* 4: 465-75.

Velasquez, J. C., Goeden, N., Bonnin, A. (2013). Placental serotonin: implications for the developmental effects of SSRIs and maternal depression. *Front. Cell. Neurosci.* 7:47.

Walker, M. B., Miller, C. T., Swartz, M. E., Eberhart, J. K., Kimmel, C. B. (2007). Phospholipase C, beta 3 is required for Endothelin1 regulation of pharyngeal arch patterning in zebrafish. *Dev. Biol.* 304, 194-207.

Wheeler, G. N., Brändli, A. W. (2009) Simple vertebrate models for chemical genetics and drug discovery screens: lessons from zebrafish and Xenopus. *Dev. Dyn.* 238, 1287-308.

Whitaker-Azmitia, P. M. (1999). The discovery of serotonin and its role in neuroscience. *Neuropsychopharmacology* 21, 2S-8S.

Whitaker-Azmitia, P. M. (2001). Serotonin and brain development: role in human developmental diseases. *Brain Res. Bull.* 56, 479-485.

Yoshiba, S., Hamada, H. (2014). Roles of cilia, fluid flow, and Ca2+ signaling in breaking of left-right symmetry. *Trends Genet.* 30,10-7.

Zhang, S., Li, J., Lea, R., Amaya, E., Dorey, K. (2013). A functional genome-wide in vivo screen identifies new regulators of signalling pathways during early Xenopus embryogenesis. *PLoS One* 8(11):e79469. eCollection 2013.

In: New Developments in Serotonin Research
Editor: Ming D. Li
ISBN: 978-1-63482-220-6

Chapter 5

Effect of Serotonin on Exotrophy Processes in Fish

Victoria V. Kuz'mina [*]
Papanin Institute for Biology of Inland Waters, Russian Academy of Sciences, Borok, Yaroslavl Oblast, Russia

Abstract

On the basis of original studies and the data by published other authors a role of serotonin in the regulation of processes in exotrophy in fish is analyzed. The data on localization and functions of serotonin in mammals and fish are given along with information on the content of this hormone in the fish of the same species and in intestines of various fish species. A special attention is paid to the role of serotonin in the regulation of fish foraging behavior and in the processes of digestion, effects on the activities of digestive hydrolases in particular.

It is shown that both at the central (intraventricular) and peripheral (intraperitoneal and intramuscular) administrations serotonin affect feeding behavior in fish. During first few hours following serotonin administrations the amount of food consumes and speed of locomotion in fish may decrease. At the same time the activities of proteases and glycosidases and the level of glycemia increase.

---

[*] Corresponding author: Victoria V. Kuz'mina, Papanin Institute for Biology of Inland Waters RAS, 152742, Borok, Yaroslavl, Russia. Tel.: (48547) 24526, fax (48547) 24042, E-mail address: vkusmina@ibiw.yaroslavl.ru.

The mechanisms of direct and indirect effects of serotonin on the processes of exotrophy in fish are discussed with respect to participation of various hormones and neurotransmitters and nervous system.

Keywords: serotonin, fish, exotrophy, feeding behavior, digestion, metabolism, regulation

Introduction

Exotrophy represents a set of processes providing assimilation of food items. Exotrophy consists of a series of steps: capture of food, feeding and digestion of ingested food, absorption and transport of nutrient. Consumption of food is one of the most important features of a living being and it includes several behavioral reactions. The problems related to the mechanisms of formation of hunger and satiety have been the focus of much research for a long time. The principal scheme describing the regulation of appetite was proposed by I. P. Pavlov in 1932 (1951) who formulated the concept of "feeding center". According to Pavlov, "feeding center" is a functional system that includes the structures located at various levels of central nervous system. This center receives chemical signals ("hungry blood") and based upon these signals organizes the behavior directed to consumption of food and secretion of digestive glands necessary to food processing. Later on the Pavlov's scheme of feeding center was substantiated by the results of numerous studies that formed a base for several theories of appetite (glycostatic, aminoacidstatic, lipostatic, thermostatic, hydrostatic and metabolic). The majority of these theories were based on the consideration of the role of only a single factor. Above mentioned approaches to the understanding of the processes of self-regulation (based on the concept of C. Bernard on the constancy of internal environment which was considered as a necessary condition of "free life") were focused only on the internal organismal processes regulating feeding behavior. According to these concepts the starvation of tissues was believed to be a primary reason for the urge to food consumption (Anokhin, 1975).

The studies of mammal revealed that the "feeding center" is localized in the hypothalamus (Anand, Brobeck, 1951). Later on it was shown that the destruction of ventromedial zone of hypothalamus leads to the hyperphagy; of the lateral zone, to the refuse for feeding. It was considered that the former zone is a center of satiety; the latter, the center of hunger. In addition, it was proved that numerous neurotransmitters and hormones participate in the regulation of feeding behavior in vertebrates, including fish. It is known that noradrenalin, γ-aminobutyric acid, galanin and melanin-concentrating hormone stimulate food consumption, while serotonin, dopamine, histamine, corticoliberin, cholecystokinin, α-melanocyte stimulating hormone, YY peptide, bombesin and other peptides and amines inhibit motivation to feed. The growth hormone and the functional "axis" of growth hormone / insulin-like growth factor play important role in the regulation of food consumption. Above mentioned and some other facts assume that feeding behavior in fish, as well as in the higher vertebrates is regulated by the multichannel system that includes all known mechanisms of nervous and humoral control. The integration of these mechanisms is provided in hypothalamus with participation of various structures of brain based on the signals from the external and internal environments of an organism (Kuz'mina, 2005).

Some attempts to assess the role of diverse factors in the regulation of the complex of reactions ensuring the processes of exotrophy in fish have been recently made. On the basis of the analysis of neurohumoral mechanisms controlling feeding behavior structure, feeding, digestion and other links of exotrophy as well as the processes of endotrophy and metabolism in fish, the assumption that a unified outline (contour) of exo-, endotrophy and metabolism processes regulation exists has been formulated (Kuz'mina, 2000, 2005).

It was partially confirmed at the analysis the role of gastroenteropancreatic hormones in the regulation of metabolism in fish (Nelson, Sheridan, 2006). It was show that food intake is promoted by galanin, neuropeptide Y, and pancreatic polypeptide, while cholecystokinin and glucagon-like peptide-1 inhibit food intake. Digestion of ingested food is facilitated by cholecystokinin, pancreatic polypeptide and secretin by coordinating gastrointestinal tract motility and regulation of exocrine secretion. Somatostatins, on the other hand, generally inhibit exocrine secretions. Insulin facilitates assimilation by promoting the uptake of nutrient molecules (e.g., glucose, aminoacids, and fatty acids) into cells. Insulin also is generally anabolic and stimulates the synthesis and deposition of energy reserves (e.g., glycogen, triacylglycerol) as well as of proteins, thereby facilitating organismal growth. Insulin-like growth factors (e.g., IGF-1) also promote cell proliferation and organismal growth. Breakdown and mobilization of stored energy reserves is stimulated by glucagon, glucagon-like peptide-1, and somatostatin. Somatostatins also affect metabolism and reproduction via their effects on the thyroid axis as well as growth via effects on growth hormone release and perhaps directly via modulation of growth hormone sensitivity.

Regarding the analysis of the coordinative role of hormones and neurotransmitters in the regulation of exotrophy processes, the investigation of serotonin effects is of particular importance. Serotonin (5-hydroxytryptamine or 5-HT) is one of the most ubiquitous neurotransmitters in the animal kingdom (Azmitia 1999; Barry, 2013). It is generally accepted that 5-HT is produced from tryptophan. Besides, a complex model to explain the relationship between circulating tryptophan and central 5-HT has also been proposed. In this scheme, brain 5-HT synthesis is dependent not only on levels of circulating tryptophan, but also on levels of other long neutral amino acids (LNAA) that compete with tryptophan for transport into the brain. High tryptophan/LNAA ratios promote central 5-HT synthesis, while low tryptophan:LNAA ratios have the opposite effect. Protein-rich meals decreases this ratio, while carbohydrate-rich meals increases it, presumably by stimulating insulin release, which in turn promoted absorption of LNAAs by peripheral tissues (Lyons, Truswell, 1988). Brain 5-HT is synthesized in neurons where it is stored in dense vesicles on the presynaptic membrane, whereas peripheral serotonin is synthesised and released by enterochromaffin cells in gut mucosa in response to different stimulus such as gastric distension, hyperosmotic solutions or presence of glucose (Rubio et al., 2006).

It is important to note that 5-HT is synthesized in various tissues but most presents at the highest amount in the intestine. This suggests its important role not only in the central regulation, but also in the peripheral regulation of exotrophy processes. However despite the considerable relevance of the problem these aspects of feeding behavior of fishes remain poorly studied to date. At the same time the study on the regulation of exotrophy in fish clarifies the evolution of the regulatory mechanisms and proves the data necessary for improvement of the fish farming techniques.

In the present chapter up-to-date data concerning the role of the hormonal system serotonin in particular, in the regulation of exotrophy in fish are discussed. Special attention will be paid to the role of serotonin in the regulation of feeding behavior and digestive processes in fish.

1. Localization of Serotonin

Several studies revealed that synthesis of 5-HT takes place in various tissues of vertebrates. However the data on its localization in brain and digestive tract are most important for the analysis of its role in the processes of exotrophy.

In the mammals'CNS serotonin is synthesized exclusively in discrete assemblages of brain stem neurons (the *raphe nuclei*). The caudal *raphe nuclei* send predominantly descending projections to the brainstem and spinal cord (Lam and Heisler, 2007). The rostral *raphe nuclei*, including the medial and dorsal *raph*e, send predominantly ascending projections throughout the forebrain, including to the cortex, amygdala, striatum, hippocampus, and the various nuclei of the hypothalamus. The most areas of the hypothalamus receive inputs both from the medial and dorsal *raphe* (Petrov et al., 1992). The studies using the immune-chemical techniques revealed that the brains of cartilaginous (Ritche et al., 1983) and bony (Fontaine, 1975) fishes contain well-developed serotoninergic system. General structures of ascending and descending cerebral pathways in fish and mammals are similar to each other (Kah, Chambolle, 1983). Neurons using 5-HT as neurotransmitter and/or modulator have been identified in the central nervous system in representatives of jawless, cartilaginous and ray-finned fishes (Lillesaar, 2011).

Various tissues of mammals generate peripheral 5-HT produced in a range of tissues. In platelets and in the gastrointestinal tract 95% of 5-HT is localized (Essman, 1978). In the intestine, 5-HT is located along the entire length, both in neurons and in the enterochromaffin cells of the intestinal epithelium. This compound is released initially in the underlying layer of connective tissue of the mucosa, which are innervated by the nerve endings of primary sensory neurons in the enteric nervous system. The high concentration of 5-HT flowing in the circulatory system and the intestinal lumen are revealed (Donovan, Tecott, 2013). It is known that 5-HT exerts its effect, with the participation of at least 18 different receptors of 5-HT (Marston et al., 2011).

In fish 5-HT was found in various tissues as well. It is important that in fish 5-HT is produced by enterochromaffin cells localized in the intestine epithelium. Intestine is one of the main peripheral sources of 5-HT in fish (Piomelli, Tota, 1983; Caamaño-Tubío et al., 2007). This compound appears at the early stages of fish evolution. For example, the intraepithelial cells immune-responsive to serotonin were revealed in the intestine of cartilaginous fish (shark *Squalus acanthias*).

In addition, the presence of serotonin-containing cells was proved using the methods of fluorescent histochemistry (Holmgren, Nilsson, 1983). The contents of 5-HT vary in the intestines of different fish species (Table 1). At that, the level of 5-HT not necessarily correlates with the peculiar features of fish biology and feeding modes. At a first glimpse it appears that the level of 5-HT is much higher in the typically piscivorous pike than in the typically benthivorous carp and whitefish.

Table 1. The level of serotonin in the intestine of various fish species (Terenina, Gustafsson, 2003)

Fish species	Level of 5-HT, μg/g of tissue
Pike *Esox lucius*	1.400 ± 0.030 (7)
Carp *Cyprinus carpio*	0.621 ± 0.092 (5)
Salmon *Salmosalar*	0.289 ± 0.043 (9)
White fish *Coregonus lavaretus*	0.210 ± 0.064 (3)
Burbot *Lota lota*	0.121 ± 0.024 (13)

Means ±S.E. are given. Number of fish examined is given in parentheses.

However, the differences in the levels of 5-HT between pike, salmon and burbot (the representatives of the groups of piscivorous and piscivores-facultative benthivores) are more pronounced.

In addition to species-specific differences the same authors revealed significant changes in the content of 5-HT in the intestine during the annual cycle, indicating that its level depends on the intensity of fish feeding (Table 2).

Table 2. Influence of the season on the level of serotonin in the intestine of carp (Terenina, Gustafsson, 2003)

June-July	August	September	October	December
1.050±0.044 (22)	0.621 ± 0.092 (5)	0.340±0.059 (5)	0.286±0.024 (4)	0.286±0.025 (4)

Means ±S.E. are given. Number of fish examined is given in parentheses.

Maximal level of 5-HT was recorded in June-July, the months of most active feeding in carp. Later, in August and September this level gradually decreases which coincides with decreasing intensity of carp feeding.

Finally, in the late fall and winter when the exogenous feeding is much less intensive or ceased the level of 5-HT decreases by 3.7 times comparing to the maximal value. In this case the changes in temperature may play an important role. For instance, the rise in temperature from 14°C to 19°C leads to almost 20 % increase in the level of 5-HT in the brain of eel *Anguilla anguilla* (Sebert et al., 1985).

The contents of 5-HT revealed in various fish tissues were also considerable. In rainbow trout (*Oncorhynchus mykiss*) the organ levels of 5-HT (expressed as ng/g of tissue) were revealed in the following order: anterior intestine (2624±327)> pyloric caeca (2272 ± 238) > the middle + posterior intestine (1799 ± 566) > gill epithelium (1569±440)> anterior kidney (160±26) > posterior kidney (116±38) >liver (28.5±8.5) >heart (23±6) > spleen (0.22±0.05). 5-HT levels observed in blood did not differ significantly from those observed in plasma 1173±407 and 807±99 pg/mL, respectively (Caamaño-Tubío et al., 2007). Considerable differences in the level of 5-HT in the intestine of the fish of the same fam. Salmonidae (salmon and trout) are also note worthy. Presumably these differences relate to the different time periods when these fish were studied and, hence, to the intensities of their feeding. The content of 5-HT in the brain of juvenile Atlantic cod *Gadus morhua* is low as well: 167±28 ng/g. It was shown that in trout high intestinal 5-HT concentration is mostly associated with serotoninergic nerve fibres, present at high density in the intestinal wall (Caamaño-Tubío et al., 2007). In rainbow trout only about 2% of 5-HT is associated with mucosal enterochromaffin cells. The most part of 5-HT is connected not to the enterochromafine cells of a mucosa, but to the serotoninergic nervous fibres (to 98 %), densely located in an intestinal wall. In addition, as it was shown in this paper 5-HT released from intestinal serotoninergic fibres in response to d-fenfluramine treatment is metabolized locally, and only a small part reaches the blood, from where it can be taken up and metabolized by other peripheral tissues, such as the liver tissue. It is assumed that non-metabolized serotonin pool in the blood is located extracellularly, but not intracellularly as in mammals. However, the structures of peripheral nervous systems may vary in the fish of different taxa. For example, in some bony fishes (*Mugil auratus* and *Salmo trutta*) the serotonin-containing neurons are found in the intestine while in other species (*Anguilla anguilla* and *Platessa platessa*) no such neurons were found (Anderson, 1983).

2. The Functions of Serotonin

Vertebrate and invertebrate species utilize serotonin for the regulation of many of the behavioral and physiological processes as well as for the metabolism. In particular, in many animals, 5-HT plays an important role in the regulation of appetite (Anand, Brobeck, 1951;Winberg et al., 1993b; Simansky, 1996; Hoebel, 1997; de Pedro et al., 1998; Haider, Haleem, 2000; Rubio et al., 2006), feeding behaviour (Leibowitz, 1986), feeding motivation (Silva et al., 2014), thermoregulation (Rothwell, Stock, 1987; Tsai et al., 2002; Bicego et al., 2007), energy balance (Marstonet al., 2011; Donovan, Tecott,

2013), metabolism (Yamada et al., 1989; Watanabe et al., 2010a), locomotor activity (Winberg et al., 1993a), establishment of social behaviour and hierarchies (Overli et al., 1998; Larson, Summers, 2001), aggressive interactions (de Almeida et al., 2005; Lepage et al., 2005; Höglund et al. 2005; Zubizarreta et al., 2012), cardio-vascular system (Janvier, 1997; Forster et al., 1998) and stress response (Emerson et al., 2000; Barton et al., 2002; Leonard, 2005; Kumar et al., 2014) as well as an array of other functions.

In mammals, the level of hypothalamic 5-HT increases under feeding and decreases under food deprivation (Schwartz et al., 1989; Haider, Halim, 2000). The lesion of medial hypothalamic (ventromedial, dorsomedial nucleus, paraventricular) nuclei leads to hyperphagia and obesity, while malfunction of the lateral hypothalamus causes hypophagia (Anand, Brobeck, 1951).

Over the past few years, identification of a number of orexigenic and anorexigenic signaling molecules and their receptors in the hypothalamus, and the neuronal sites of their production, release, and receptive fields has redefined the hypothalamic pathways involved in the regulation of feeding and body weight. In addition, evidence of morphological relationships among these signaling-molecule producing neurons, and co-localization or co-production of more than one signaling molecule in these neurons has demonstrated an interconnected circuitry in the hypothalamus to regulate feeding behavior (Lin et al., 2000).

Importantly, as it was shown central injection of 5-HT or 5-HT precursor 5-hydroxytryptaphane cause hypophagia (Simansky, 1996; Yamada et al., 2006; Watanabe et al., 2010b) and stimulates energy balance (Watanabe et al., 2010a; Donovan, Tecott, 2013). Increase in the level of 5-HT in hypothalamus results in the decrease in body temperature (Bicego et al., 2007). Regulatory effect of 5-HT on the feeding is mediated by 5-HT receptors located in various nuclei of the hypothalamus (Leibowitz and Alexander, 1998). Regulatory effect of 5-HT on the feeding mediated by 5-HT receptors located in various nuclei of the hypothalamus (Leibowitz and Alexander, 1998). It is known that 5-HT exerts its effect, with the participation of at least 18 different 5-HT receptors (Marston et al., 2011). At the same time, a considerable amount of 5-HT is located in the periphery. The study on the endo- and paracrine role of 5-HT in the regulation of digestive system in mammals revealed that 5-HT acts as a peripheral mediator of satiety, and inhibitory effect of 5-HT is realized through two types of receptors – 5-HT_{1B} and 5-HT_{2C} (Simansky, 1996). The releasing of 5-HT by enterochromaffin cells of the digestive tract is stimulated in the presence of food (Fujimiya et al, 1997).

Serotonin participates in the functioning of the peripheral part of reflectory arch. In mammals, at low doses (below 50 μg / kg body weight) 5-HT modulates synaptic transmission in the sympathetic ganglia and in the vegetative nerve-muscular plexuses. At the doses of about 100 μg / kg body weight this compound facilitates the synaptic transmission in the synaptic ganglia; at 660-700 μg / kg body weight, inhibits functions of synaptic structures (Nozdrachev, 1981). Direct impact of 5-HT on digestive tract is specific and relates neither to the effects of acetylcholine nor of histamine. That is, 5-HT is not a mediator of cholinergic excitation of motor visceral innervation (Shuba, 1981). In fish 5-HT enhances the stomach and intestine peristaltic (Schparkovsky, 1986). In addition, 5-HT mediates an animal's behavioral response to social stress (Stoddard et al., 2008).

The predominant effect of central nervous system serotonin signaling is the suppression of food intake. Central administrations of serotonin induce hypophagia (Simansky, 1996; Yamada et al., 2006) and increases metabolic rate (Rothwell, Stock, 1987; Le Feuvre et al., 1991). In fish, 5-HT seems to play a role in integrating behavioral and neuroendocrine stress responses (Overli et al., 1998), but no effects on energy metabolism have been revealed to date (Soengas, Aldegunde, 2002).

3. Regulation of Feeding Behavior

Traditionally, significant attention is paid to the central regulation of feeding behavior in mammals and fish (Schwartz et al., 1989; Hoebel, 1997; Lin et al., 2000; de Pedro, Bjornsson, 2001). The serotonergic system is one of the intermediary systems involved in such control in fish (Bernier, Peter 2001; de Pedro, Bjornsson, 2001). At that 5-HT is considered to be one of the main inhibitors of foraging behavior in fish (Figure 1).

This scheme shows that the digestive tract hormones are regulators of feeding behavior. However, 5-HT synthesized in the intestine does not shown to participate here. At the same time it was revealed that low serotonergic activity in the hypothalamus was highly correlated with fish individual features characterized by high feeding motivation and feeding behaviour, namely feeding latency and feeding score (Silva et al., 2014).

Furthermore, in the study of fish the possibility of involvement of serotonin in the central regulation of appetite was also demonstrated. The intracerebroventricular injections of 5-HT (10 μl/g body weight) induce a significant decrease in food intake in goldfish *Carassius auratus*.

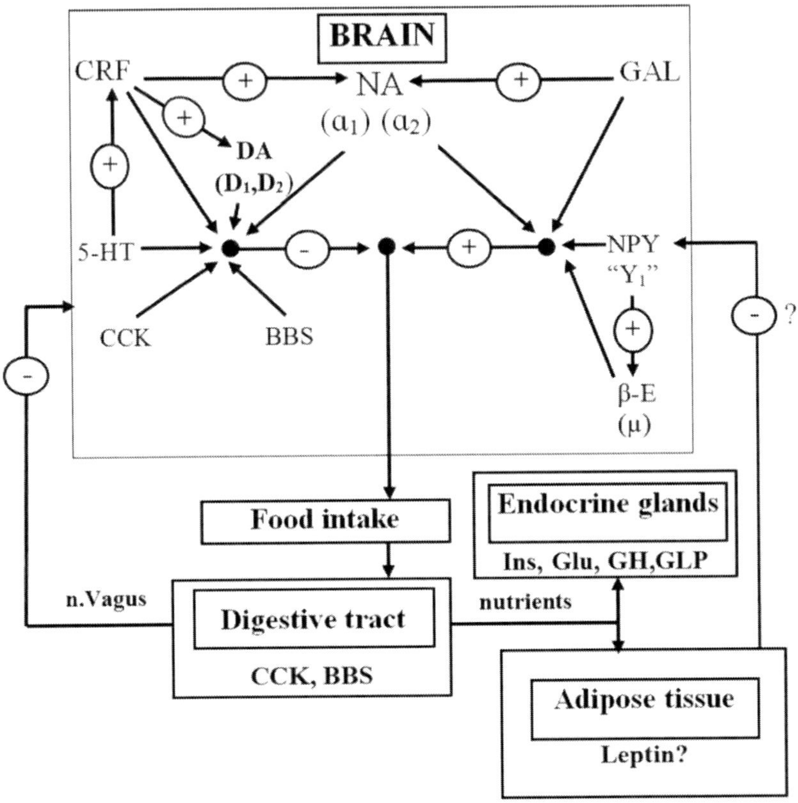

Legend: CRF – corticotropin-releasing factor, NA – noradrenaline, GAL – galanin, DA – dopamine, 5-HT – serotonin, NPY – neuropeptideY, CCK – cholecystokinin, BBS – bombesin, Ins – insulin, Glu – glucagon, GH – growth hormone, GLP – glucagon-like peptide, n. Vagus – the vagus nerve, α_1, α_2 – adrenergic receptors, D_1, D_2 – dopaminergic receptors, Y_1 – receptors of neuropeptideY, β-E – β-endorphin, μ – opioid receptors.

Figure 1. Schematic diagram of regulation of food consumption by neuropeptides and hormones in fish (de Pedro, Bjornsson, 2001).

This inhibitory effect on appetite is shown at short time (2 h after 5-HT administration), but not during the next discrete intervals (2–5 and 5– 8 h). After the peripheral administration of 5-HT (1 and 10 μg/g body weight), no significant differences in food intake were detected at any of the analyzed intervals: 0 –2 h. Cumulative food intake at 8 h after 5-HT administration using both techniques did not show differences in 5-HT-treated with respect to the control fish (de Pedro et al., 1998).

In this work was suggested that the decreasing of food intake by fishes after intracerebroventricular injection of 5-HT can be conditioned by participation of CRF which caused not only direct anorectic effect but also mediated the effects of 5-HT. The results of the experiments with administration of antagonists of CRF-receptors gave evidence that CRF can be involved in 5-HT effects by activation of CRF-neurons which in one's turn inhibit serotoninergic transmission (de Pedro et al., 1998; Lowry, Moore, 2006). Taking in mind the anoretic effect of CRF in goldfish the authors suggested that this neuropeptide can act as mediator in the 5-HT anoretic action. These facts suggest the participation of serotonin in central anoretic effects only (de Pedro et al., 1998).

At the same time the possibility 5-HT molecule transfer from the fish digestive tract into the blood has been established. The administration *per os* of gelatin capsules containing 5-HT (0.1, 0.5 and 2.5 mg/kg bw) as well as carbohydrates, protein, or lipids separately in sea bass *Dicentrarchus labrax* leads to the increase of plasma 5-HT level two times greater than the controls, 20-45 min after the ingestion of a capsule containing 2.5 mg 5-HT/kg bw. The rise of the plasma 5-HT was accompanied by a fall of the fish food intake. The indoleamine, at doses of 0.1, 0.5 and 2.5 mg kg bw, produced a reduction in total food intake of 31%, 49% and 37%, respectively, compared to the baseline, modifying the macronutrient selection pattern. The percentage of fat was significantly reduced whereas the percentage of protein significantly increased after administration of highest dose, but no changes were observed in the proportion of carbohydrate for any 5-HT doses (Rubio et al., 2006).

These facts suggest 5-HT involvement in the regulation of fish feeding behavior not only through the central but also through the peripheral mechanisms.

In addition, it was shown that sublethal doses of fluoxetine (a selective 5-HT reuptake inhibitor) caused disorders in feeding behavior of the hybrid striped bass *Morone saxatilis x M. chrysops*, in particular, prey catching speed changes that correlated with the decrease in the brain's serotoninergic activity (Gaworecki, Klaine, 2008)). The intraperitoneal introduction of fenfluramine, a serotonin-releasing agent, induced a short-term inhibition of food intake in rainbow trout *Oncorhynchus mykiss*. The authors of refereed paper suggest that the increase in the content of 5-HT in the hypothalamus and the telencephalon as well as decrease in food consumption in the rainbow trout may be associated with an increase in 5-HT availability in the synaptic cleft as it is reflected in changes in the 5-HIAA/5-HT ratio. In order to make the information more comprehensive, it should be mentioned that starvation during three weeks considerably diminished its content in these brain regions (Ruibal et al., 2002).

Later it was found that intraperitoneal injection of 5-HT in the same dose (10 μg/g bw) not only reduced the amount of food consumed, but also modified the locomotor activity in common carp *Cyprinus carpio* fingerlings (Kuz'mina et al., 2010; Kuz'mina, Garina, 2013). These experiments revealed that intraperitoneal injection of serotonin (1, 5 and 53 h post treatment) significantly increases the latency to leave the 10 x 5 x 6 cm starting chamber constructed of transparent perforated walls after its front partitioning was raised enabling the fish to leave the chamber. The speed of feeding reaction or the time required to reach the food (30 frozen specimens of chironomid larvae) at the opposite wall of the aquarium, increased considerably 1 h after the injection of 5-HT. In addition, significant effect was recorded 5, 29 and 53 h following the injection of 5-HT (Figure 2).

The analysis of data concerning food intake also revealed insignificant variability of this parameter in control fishes. In the experimental fish significant decrease (- 35%) in food intake (from 27.8±0.7 to 17.9±1.3 chironomid larvae, $p \leq 0.05$) was observed 1 h after the injection. Later, 5 and 24 h after the injection, the values of this parameter were still below the control level. No significant differences between the values of food intake in the treated and control were revealed during further observations. The study on the effects of intramuscular injections of serotonin on fish feeding behavior indicate principal similarity of serotonin effects on fish feeding reactions at both ways of hormone administration (Kuz'mina, Garina, 2013).

It is worth noting that the effects of peripherally administered serotonin were different in goldfish and carp. Earlier, the absence of the effect of intraperitoneal injections was explained by very low capability of molecules of 5-HT to cross through the blood-brain barrier. In addition to the species-specific features, above differences may be due to different illumination regimes. In the study by de Pedro et al., 1998 the experiments were carried out 12 h: 12 h light/dark regime; in the studies given in (Kuz'mina et al., 2010; Kuz'mina, Garina, 2013), light/dark cycling was as follows: 8 hs "light", 450 lux and 16 hours "dark", 0.08 lux. It possible that in the conditions of weak lightening the level of endogenous 5-HT in fish organisms was lower than at lighter experimental environment. It is known that illumination level of 430-2200 lux is optimal for growth of carp (Ruchin, 2001). Apparently, low content of endogenous 5-HT in fish bodies can lead to stronger effect of exogenous 5-HT on feeding behavior in carp. The involvement of adrenaline in the regulatory process is possible as well. Actually, the stimulation of intestinal motility following 5-HT injection can be conditioned by adrenaline release since it is known that serotoninergic system and hypothalamo-pituitary-adrenal axis in fish are closely cooperative (Bernier, Peter, 2001).

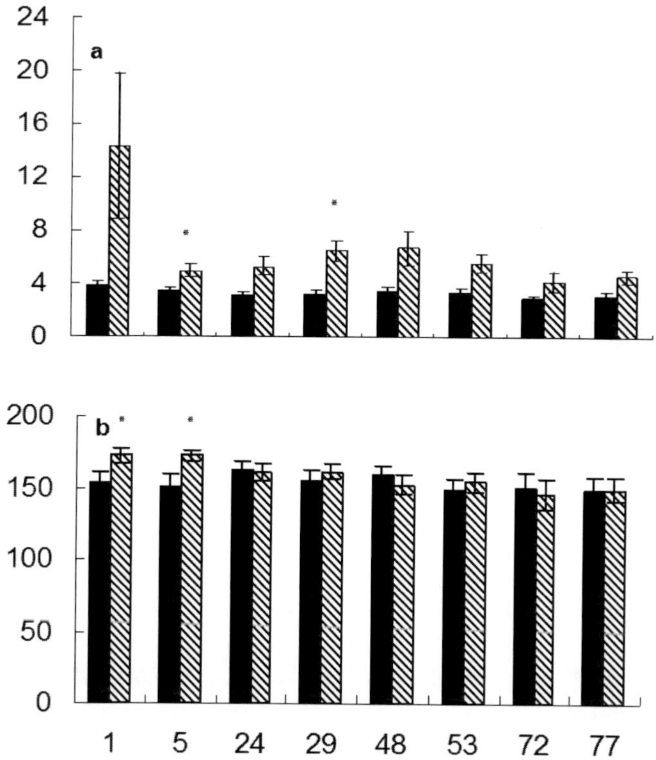

Legend: Abscissa: time after the injections, hours; ordinate: a – time characteristics of feeding behavior, s., b – the number of chironomid larvae consumed for 3 min. Dark bars – control fishes, hatched bars – experimental fishes. Means ± SEM, are given. *P <0.05.

Figure 2. The effects of intraperitoneal injections of serotonin on speed of feeding reaction (a) and food intake (b) in the experiments with single feeding of carp yearlings (Kuz'mina et al., 2010).

Very interesting results were obtained for the comparison of data relating to the effects of serotonin on the duration of “single” feeding, the duration of “group” feeding, total duration of feeding and food intake (Table 3).

Table 3. Parameters of feeding behavior of carps under the administration of serotonin in conditions of group feeding, all time of observation (Kuz'mina, Garina, 2013)

Injected substances	Parameters of feeding behavior				
	t_s	t_g	t_t	R	S
5-HT (10 μg/g), IP	95±14 (53±25)	88±24 (135±44)	183±28 (187±19)	21±1.2 (23±1.1)	106±8 (76±5)
5-HT (10 μg/g), IM	40±13 (111±52)	10±3.3* (114±27)	51±14* (225±63)	10±1.7* (24±0.8)	27±5* (79±6)
Ringer's solution, IP	119±17 (115±22)	168±15 (228±24)	287±25 (342±46)	25±0.7 (25±0.7)	107±11* (160±4)
Ringer's solution, IM	123±21 (161±24)	133±26 (179±17)	256±43 (340±37)	20±2.1 (24±0.7)	76±6 (118±1)

Note: Values are the mean±S.E (n=9 for each behavioral parameter and time interval); t_s – duration of "single" feeding, s., t_g – duration of "group" feeding, s., t_t – total duration of feeding, duration of "digging" and pinching of food particles including both duration of "single" (only one fish, t_s) and "group" (two or three fish simultaneously, t_g) feeding, s., R – food intake, total amount of food consumed by a group of fishes during the observation time (10 min), chironomid larvae, S – locomotor activity, quantity of crossings of two vertical lines on the mirrow which dividing surface area of the aquarium into three equal parts, by individual fishes swimming above the bottom. *significant differences from "intact" values. In brackets the values of parameters of intact fishes (for each experimental group) are indicate.

This experiment allows reveal the features of "group" behavior of fishes. Behavior of intact carps looks so: after the removing of starting chamber carps began to investigate near-bottom water layers of the aquarium, swimming quickly and digging sand in different places. When a fish found a "food spot" other carps immediately joined to it. The fishes didn't dig ground on "food spot" for a long time. They pecked it sometimes in this place and then left to investigate bottom further. Then digging sand on "food spot" became longer. Behavior of carps injected with serotonin significantly differed from one of intact specimens. After leaving the starting chamber fishes began swimming together but the school disappeared quickly (after 2-3 min), and carps moved and searched food separately one from another. Fishes often swim in middle and upper water layers.

The strongest effect was registered for "group" feeding of carp fingerlings. Maximal effect was observed in 1 h after injections. This parameter decreased significantly during all period of observation under intramusculary injections, and only in 1 h after the intraperitoneal injections. Significant decreasing of total duration of feeding after intramusculary injections was revealed during all period of observation whereas after intraperitoneal injections only tendency to lowering was observed. The analysis of locomotor activity of carps also demonstrated the decreasing of the parameter value under intramusculary injection in comparison with intraperitoneal administration. Maximal effect was observed in 24 h after the treatment both intraperitoneal and intramusculary injections of 5-HT lowered the food intake of carps (Kuz'mina, Garina, 2013).

These experiments for the first time demonstrated significant decreasing of food intake and locomotor activity of the carps both under intraperitoneal and intramuscular administration of 5-HT. Coincidence of time intervals of fish moving into upper water layers after the injection of 5-HT and the period of maximal concentration of 5-HT in blood of European see bass *Dicentrarchus labrax* after *per os* administration (Rubio et al., 2006) allow suggest the possibility of the fast delivery of monoamine into blood in these experiments. The moving of carps into upper layers of water appears to be conditioned by vasoconstriction of branchial vessels (Forster et al., 1998) which lead to worsening of respiration and surfacing of fishes.

It is important that intramuscular injections of serotonin cause more pronounced effects in the food-foraging behavior and motor performance in carps than intraperitoneal injections; the same is observed in mammals (Kaya et al., 1992). It is suggested that stronger effect of intramusculary injections in comparison with intraperitoneal injections of 5-HT is conditioned by higher rate of delivery of 5-HT from muscle into blood whereas peritoneum delays inflow of 5-HT into blood. The decreasing of locomotor activity of carp correlates with data testified to lowering of spontaneous locomotor activity and investigatory behavior of arctic charr *Salvelinus alpinus* after pharmacological stimulation of brain serotoninergic activity (Winberg et al., 1993b).

Thus, the inhibitory effects of serotonin on food feeding and motor performance may be preconditioned by the processes at central nervous systems as well as the peripheral nervous systems. At the same time there are the evidence of endocrine and paracrine role of 5-HT on gastrointestinal tract and vascular system indicate that 5-HT acts as a peripheral mediator of satiety (Simansky, 1996). According to de Pedro et al. (1998) these results can be due to indirect effects of 5-HT, such as actions on gastrointestinal tract (e.g., as a local modulator of motility; or even to an unspecific behavioral effect of 5-

HT). Taking in mind the anoretic effect of CRF in goldfish they hypothesis is that this neuropeptide can act as mediator in the 5-HT anoretic action.

4. Regulation of Digestive Processes

Digestion is a complex process, involving enzyme and fluid secretions, as well as motility, culminating in absorption and evacuation. The available data on hormonal regulation of the gastrointestinal tract in fish suggest that the regulatory pathways and molecules are conserved among vertebrates, but specific responses in fish may differ from mammals (Buddington and Krogdahl, 2004). Some of the gastrointestinal hormones (e.g., ghrelin, peptide YY and cholecystokinin) not only regulate digestion but may also act as appetite/satiety modulating signals in the brain (Moran, Kinzig, 2004; Volkoff et al., 2005). Hormonal regulation plays an essential role in modulating the functions of fish gastrointestinal tract to match changes in functional demands. Hormonal regulation of the fish gastrointestinal tract is described in detail by Buddington, Krogdahl (2004). They emphasize that hormones produced by the gastrointestinal tract, the associated organs (e.g., pancreas), and other sources (e.g., hypothalamus, adrenal cortex, thyroid, gonads) modulate the digestive processes (motility, secretion, and nutrient absorption) in response to dietary inputs (Table 4).

In this table 5-HT is listed as one of the signaling molecules regulating the functions of gastrointestinal tract. However no one paper proving the participation of this compound in the regulation of digestion in fish and mammals is given in the chapter “Signaling molecules that influence the GIT”.

Table 4. The different classes of signaling molecules that modulate GIT functions offish, with selected representatives for each class (Buddington, Krogdahl, 2004)

Class	Representatives
Related Peptide Families (b ~60 AA)	Gastrin-Cholecystokinin; Vasoactive intestinal polypeptide (VIP, secretin, glucagon and related peptides, gastric inhibitory polypeptide); Somatostatin; Tachykinins (Substance P); Opioids (enkephalins); Pancreatic Polypeptide (PP, neuropeptide Y, peptide YY); Others (atrial natriuretic peptide, angiotensins, gastrin releasing peptide)

Class	Representatives
Larger proteins	Growth hormone, prolactin
Iodinated proteins	Thyroxine, tri-iodothyronine
Steroids	Cortisol, estrogen, progesterone, testosterone
Prostaglandins	Prostaglandin F and prostaglandin E
Cytokines	Interleukin 1 (IL-1)
Neurotransmitters	Serotonin (5-HT), histamine

Some available data on the effect of 5-HT on the motility of fish digestive tract confirm the above suggestion. The studies on cartilaginous and bony fishes revealed that 5-HT at the doses of 20-50 μg / kg enhances the peristaltic in stomach and intestine. For instance, activation of motoric and electric activities in the stomach of ray *Raja clavata* starts 20 min following administration of this compounds and peaks following 30 min. In cod *Gadus morhua* activation of stomach motoric electric activities starts earlier (10 to 20 min following administration) but peaks later (2 h after administration) (Schparkovsky, 1986).

It is also known that 5-HT affects the activities of fish digestive enzymes. For example, in carp 5-HT (10 μg / g body weight) affects the proteolytic and amylolytic activities in the fish intestinal mucosa (Figure 3). As shown in Figure 3a, under the impact of serotonin the level of proteolytic activity changes in a wave-like manner during the experiment. At the same time, in the majority of cases (8 out of 11), there are no significant differences between the experimental and the control groups.

As it is shown in Figure 3 a, serotonin induces the "wave-like" fluctuations of level of proteolytic activity during whole span of the experiment. At the same time, in the majority of cases (8 out of 11), the differences between the experimental and control fish were insignificant. It is worth noting, that the level of proteolytic activity in the treated fish was significantly higher than in control not only on 1st and 5th hours, but on 24th and 72nd hours as well.

However, during the period of 96 to 168 hours the level of proteolytic activity sharply decreased both in the control and in the experimental fish. During this period no significant differences between enzyme activities in control and treated fish were revealed. Unlike that of proteolytic activity, the amylolytic activity dynamics in the same control group and the experimental one does not always coincide (Figure 3b). The amylolytic activity in fish of the experimental group in 7 cases out of 11 is significantly less than that of the control group.

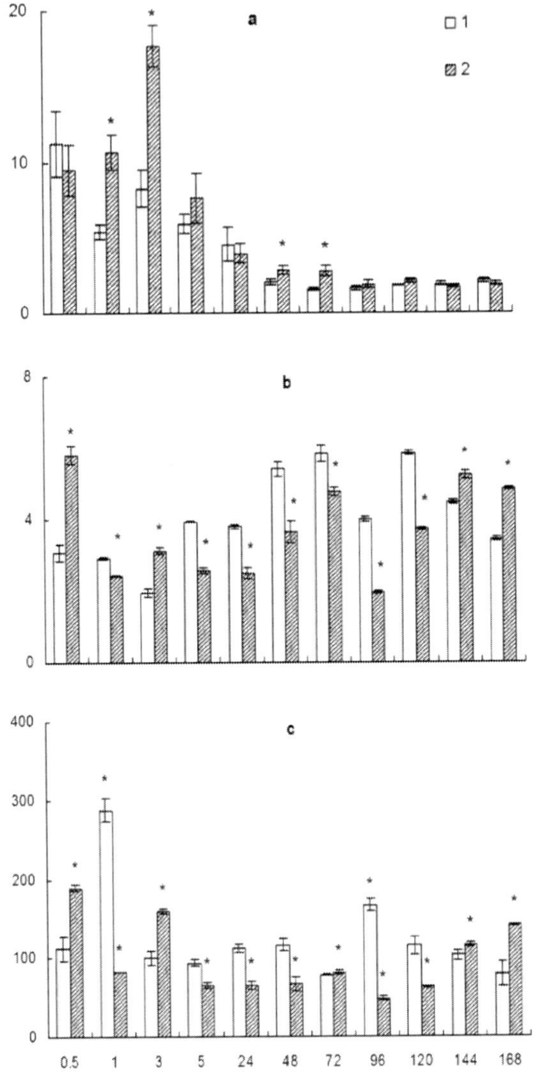

Legend: abscissa - the time after the start of the experiment, hours; ordinate – a and b: the level of enzyme activity, µmol / g min. 1 – control fish, 2 – experimental fish, c: activity of proteinase (1) and glycosidase (2) activities, % of the control group taken as 100%. *the differences are significant in comparison with the control group fish.

Figure 3. The effect of serotonin on proteolytic activity (a), amylolytic activity (b) and relative activity of proteinases and glycosidases (c) in the intestinal mucosa in carp (by: Kuz'mina, 2013).

The highest amylolytic activity values of the control group are observed in 48, 72 and 120 hours whereas those of the experimental group are observed in 72, 144 and 168 hours after the start of the experiment. It should be noted that at the end of experiment the level of amylolytic activity in fish of the experimental group is higher as compared to those in fish of the control group. As handling and injections are minor stressors, in this study the relative proteolytic and amylolytic activity has been assessed in addition (Figure 3c). The comparison of the relative activity of these enzymes shows a considerable change in their ratio in the course of the experiment. In particular, amylolytic activity surpasses proteolytic activity through 0.5, 3, 144 and 168 hours after serotonin administration (by 77, 59, 13 and 62%, accordingly).

Proteolytic activity surpasses amylolytic activity in 1, 24, 48, 96 and 120 hours after serotonin injection (by 204, 46, 49, 117 and 52%, accordingly). The maximum exceeding of the proteolytic activity values over those of amylolytic activity is recorded in an hour after serotonin injection (more than 3.5 times). The different rate of the enzymes' activity changes in the course of the experiment is important to be noted: the activity of proteinases first reverts to the norm in 3 hours after injection of the hormone whereas the activity of glycosidases – in the limits of an hour. As it has been observed, the oscillatory cycle lasts several days, suggesting involvement in the regulation of the activity of the digestive enzymes of metabolic hormones.

The rapid changes of proteolytic and amylolytic activity within the first hours of the experiment can be explained by the influence of 5-HT upon the nonstriated musculature that is responsible for contraction and relaxation of the pancreatic ducts. Indeed, it is known that 5-HT acting upon D-receptors furthers contraction of the smooth muscles (Abramets et al., 1977). Moreover, the direct and indirect action of 5-HT upon the vascular apparatus that can become apparent through vasoconstriction or vasodilatation is reported (Forster et al., 1998; Kovacs et al., 2012).

As a result, the contraction of the acinus duct walls that blocks the transit of enzymes into the pancreatic duct can occur. Relaxation of the duct walls, on the contrary, furthers the release of enzymes stocked in the pancreas' acinar cells. In the first case, enzyme activity must decrease and in the second case, these indices must increase.

Various dynamics of proteolytic and amylolytic activity suggest a different impact of 5-HT on the duct wall muscles of different acini. The subsequent changes of the enzyme activity are, most likely, caused by the alteration of trypsin and α-amylase synthesis intensity. The different degrees of dependence of these enzymes' synthesis intensity on 5-HT can be caused by the involvement of some hormones in the regulatory process.

In particular, the involvement of adrenalin in the regulatory processes is possible. It is known that the serotoninergic system and the hypothalamic-pituitary-adrenal system are closely interrelated in fish (Forster et al., 1998; Bernier, Peter, 2001). At the same time 5-HT receptors involved in regulation of pituitary-adrenocortical function (Fuller, 1986). In this regard, the observed stimulation of intestinal peristalsis after 5-HT injection can be brought about just by the release of adrenalin.

Early it was demonstrated that at low doses 5-HT or drugs that enhance the release of this neurotransmitter preferentially inhibit the ingestion of carbohydrate, more than fat or protein in mammals. This phenomenon is mediated, in part, by 5-HT receptors located in various medial hypothalamic nuclei. A negative feedback loop exists between the consumption of this macronutrient and the turnover of 5-HT in the hypothalamus. That is, carbohydrate ingestion enhances the synthesis and release of hypothalamic 5-HT, which in turn serves to control the size of carbohydrate-rich meals (Leibowitz, Alexander, 1998). In the above experiments, probably, due to the dominance of protein components in the feed for fish (proteins – 17.3 %, fats – 1.7 % and carbohydrates – 0.1 % of the raw weight) there was a decrease of amylolytic activity in almost all periods of observation, except for 0.5, 3, 144 and 168 hours. However, the increasing amylolytic activity at first period of experiment and proteolytic activity in 1 hour coincide with the significant decrease of ration and locomotive reactions under the intraperitoneal injection of the same 5-HT dose to carp fingerlings (Kuz'mina, Garina, 2013).

Moreover, it should be mentioned that the data on the nature of 5-HT action on the level of amylolytic activity are fundamentally close to those of adrenalin impact on the glycogen content in tissues (Plisetskaya, 1975). Indeed, in both cases, the oscillatory changes under adrenalin impact are observed. The similarity of the amylolytic activity dynamics and the adrenalin effect on fish feeding behavior is also noted (Kuz'mina, Garina, 2001; Kuz'mina et al., 2003). The intraperitoneal injections of adrenalin cause a multiphase increase in the latent time of fish feeding (Kuz'mina et al., 2003).

In publications on fish feeding behavior regulation, not only the interplay of the serotoninergic system and the hypothalamic-pituitary-adrenal one but also the interaction between 5-HT and corticotropin-releasing factor (CRF) is emphasized. The results obtained within the experiments on the effects of CRF-receptor antagonist introduction provide evidence of that. It has been shown that CRF can be involved in the 5-HT action through activation of the

CRF-neurons, which, in their turn, inhibit serotoninergic transmission (De Pedro et al., 1993; Lowry, Moore, 2006). Consequently, the initial amylolytic activity decrease can be caused by the action of adrenalin whereas the following – with the influence of CRF.

The oscillatory nature of the enzyme dynamics testifies the participation of metabolic hormones in regulating the activity of proteases and glycosidases which is confirmed by the dynamics of the glycemic level under serotonin administration (Figure 4 a). It was shown that the glycemic level in fish of the same experimental group initially decreases.

Then it rises, reaching a maximum in 3 h after serotonin injection. Decrease in the level of glycemia induced by 5-HT accords well with data obtained at the studies on mice (Yamada et al., 1989). It is worth noting that as it is shown in the above refereed paper 5-HT induces increase in the level of insulin.

At that time, the index values of the experimental group fish are 2.2 times higher than those of the control group fish. In 5 hours after the start of the experiment the glycemic level in fish of the both groups begins to decrease and stabilizes at a lower level until the end of the experiment. The assessment of the glycemia dynamics regarding the control group has enabled to demonstrate more clearly the oscillatory changes in this index (Figure 4 b).

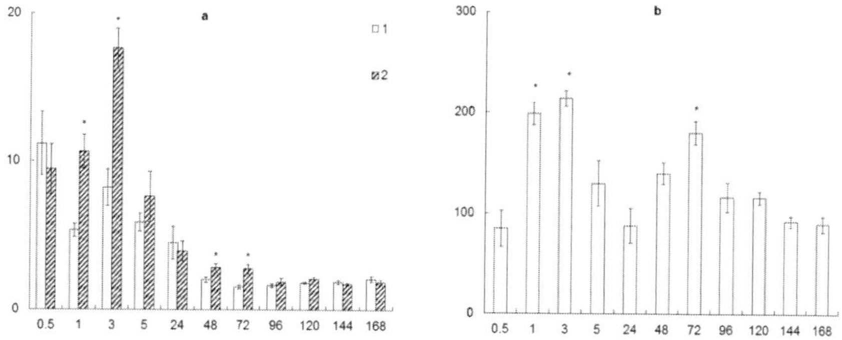

Legend: abscissa - the time after the start of the experiment, hours; ordinate – a: the level of glycemia, mmol / l. 1 – control, 2 – experimental fish, b: relative level of glycemia, % of the control group taken as 100%. *the differences are significant in comparison with the control fish.

Figure 4. The effect of serotonin upon the glycemia level in carp.

Comparison of the data given in the above figure shows that in the case of the glycemia dynamics evaluated in the absolute values, the significant rise of the index was observed only within the first three hours of the observation whereas, in the case of the relative values, in 72 hours as well. It is important to notice that the maximum glycemic level in the experimental group fish is observed in 3 hour after the start of the experiment.

The glycemic level rise is in accord with the considerable increase of this index in carps reported after the intraperitoneal injection of adrenalin (Plisetskaya, 1975). The prolonged response of the experimental group fish organism probably can be caused by both the direct reaction of adrenalin and the mediate action of 5-HT upon the hypothalamic-pituitary-adrenal system (Dinan, 1996; Bernier, Peter, 2001). In addition, the fluctuations of the studied parameters may be determined by involvement of insulin (Plisetskaya, 1975).

It was shown that intraperitoneal treatment with bovine insulin (4 mg/kg) decreased the 5-hydroxyindoleacetic acid/5-HT ratio in hypothalamus after 1 h (Rubial et al., 2002). At that the effect of insulin on feeding behavior and activity of proteases may strongly depend on the composition of food. For instance, the study of the effects of injected and dietary arginine on plasma insulin levels and growth in Pacific salmon *Oncorhynchus tshawytscha* and rainbow trout *Oncorhynchus mykiss* revealed that insulin may act as a satiety signal (Plisetskyay et al., 1991).

It is possible the different degrees of the involvement of some hormones in the regulation of metabolism and the processes of digestion that has been shown earlier within the research on the impact of insulin upon the glycemic level and the content of glycogen in fish tissues (see: Kuz'mina, 2005). These data accord well with revealed earlier increase in telencephalic levels of 5-HT after intraperitoneal glucose injection (500 mg/kg). Long-term food deprivation up to 3 weeks significantly increased hypothalamic and telencephalic levels of 5-hydroxyindoleacetic acid, whereas the ratio 5-hydroxyindoleacetic acid/5-HT significantly increased throughout the food-deprivation period (Rubial et al., 2002). Potentially the changes in the concentrations of glucose and other signal nutrients may induce the chain of hormonal reactions (Table 5).

Unfortunately as it is noted in the above the nutrients may serve as the precursors for serotonin synthesis in CNS only. At the same time glucose and other monomers formed as a result of hydrolysis of food macromolecules may trigger the release of serotonin from the endocrine cells of intestine. This serotonin in turn excites serotoninergic nerve fibres signaling to CNS.

Table 5. Proposed and potential mechanisms through which glucose, amino acids, and fatty acids might influence endocrine function in fish (MacKenzie et al., 1998)

Peripheral mechanisms
Direct stimulation of hormone synthesis and secretion (e.g., pancreatic and gut hormones); Modulation of hormone receptor availability (growth hormone receptors in liver); Alteration in post-receptor signalling in target cell (growth hormone activation of liver); Alteration of hormone transport and clearance (blood IGF binding proteins); Effects on hormone activation (liver thyroxine deiodination); Modulation of pancreatic or thyroid hormone production by nutrient-stimulated peripheral hormones (insulin effects on deiodination, CCK on insulin)
Central mechanisms (CNS and pituitary)
Direct stimulation of regulatory neuropeptide production (Neuropeptide Y); Providing precursors for neurotransmitter synthesis (serotonin, dopamine); Direct activation of pituitary hormone release (growth hormone); Modulation of hypothalamic regulation of pituitary hormone production (TSH); Nutrient-induced autonomic stimulation of hormone secretion (insulin); Modulation of hypothalamic or pituitary hormone production by nutrient-stimulated peripheral hormones (CCK and growth hormone, insulin and Neuropeptide Y)

Comparison of the data on the feeding behavior, activities of digestive enzymes and level of glycemia as affected by 5-HT indicates that, as a rule, the most alterations of studied parameters are observed during the first 24 hours after the intraperitoneal injection of serotonin (Kuz'mina, 2013; Kuz'mina, Garina, 2013).

At that decrease in the level of glycemia following 30 min post injection is accompanied by the increase in the activities of digestive hydrolases, glycosidases in particular. Also importantly that in both the experiments on the effect of 5-HT upon the fish feeding behavior and the experiments on its influence upon amylolytic activity, the inhibiting effect dominates.

The comparison of the available data also suggests that 5-HT produced by the enterochromaffin cells of intestinal epithelium (Holmgren, Nilsson, 1983; Piomelli, Tota, 1983; Caamaño-Tubío et al., 2007) influences the processes of exotrophy mainly by means of the metasympathetic nervous system.

Indeed, the major part of 5-HT is connected not with the enterochromaffin cells of the mucosa membrane but with the serotoninergic fibres (up to 98%) located in the intestinal wall (Caamaño-Tubío et al., 2007). The above assumption is confirmed by the results of the research on the endo- and paracrine role of 5-HT in the digestive system regulation in mammals indicating that serotonin acts as a peripheral satiety mediator the inhibiting effect of which on the feeding activity is realized through its interaction with receptors of two types - 5-HT_{1B} and 5-HT_{2C} (Simansky, 1996).

The obtained data make it possible to assume that 5-HT is a component of the unified outline (contour) of exo-, endotrophy and metabolism processes regulation (Kuz'mina, 2005).

According to scheme of Buddington and Krogdahl (2004) the fish gastrointestinal tract (GIT) are regulated by three hormonal signaling networks. Paracrine signals (top panel) from endocrine cells regulate the functions of nearby cells, as well as each other, such that specific functions are matched with local conditions. For intra- gastrointestinal tract signaling networks (middle panel), hormones from one region modulate the functions in other GIT regions and associated organs. Inter-organ signaling networks (bottom panel) involve axes between the gastrointestinal tract and other sources of hormones and signaling molecules (central nervous system, cardiovascular system, and the gastrointestinal tract (Figure 5).

At that paracrine signals (top panel) from endocrine cells (EC-1 and EC-2) regulate the functions of nearby cells, as well as each other, such that specific functions are matched with local conditions. For intra-GIT signaling networks (middle panel), hormones from one region modulate the functions in other GIT regions and associated organs (as indicated by arrows).

Inter-organ signaling networks (bottom panel) involve axes between the GIT and other sources of hormones and signaling molecules (central nervous system, CNS; cardiovascular system, CV) and the GIT.

The scheme given in Figure 5 suggests that 5-HT acting locally triggers transition of the signal on the entrance of nutrients into digestive tract via the receptors and serotoninergic nerve fibres included in n. *Vagus*. This signal is partially sent to pancreas triggering synthesis and translocation of digestive enzymes; partially, tp CNS, participating in the regulation of feeding behavior in fish.

That is, 5-HT not only combines regulation of such important links of exotrophy as feeding behavior and digestion, but also serves as component of the contour of regulation of exotrophy and metabolism.

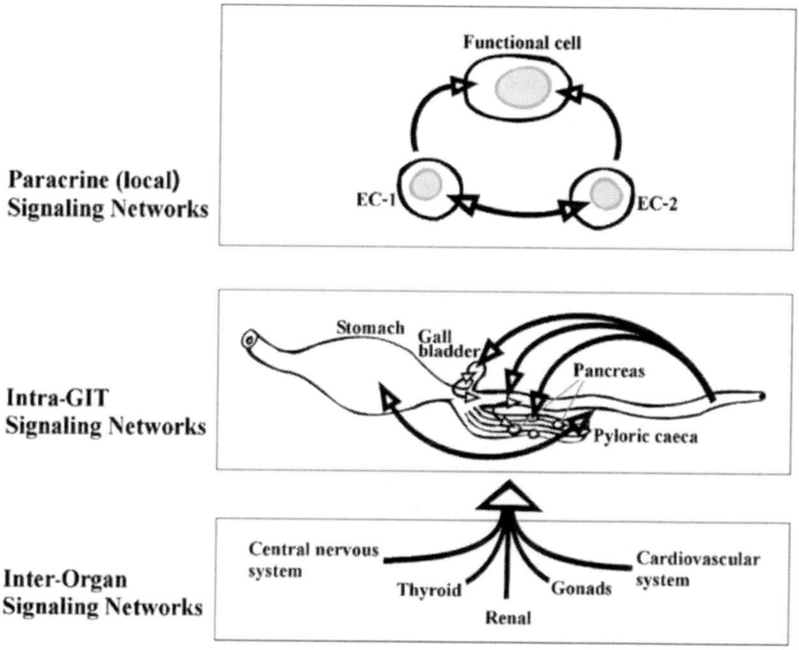

Figure 5. GIT functions are regulated by three hormonal signaling networks (Buddington, Krogdahl, 2004).

Conclusion

The importance of 5-HT signaling in the regulation of food intake, digestion and metabolism has been demonstrated. Particular attention has been paid to mechanisms through which 5-HT impacts pathways of these processes within the hypothalamus and intestine. First the data on the impact of the same 5-HT dose upon feeding behavior and the activity of digestive enzymes in the same fish species inhabiting the same environment are present. It is shown on the basis of original and literature data that 5-HT decreases food consumption both if administered centrally and peripherally. In addition it has been revealed that serotonin affects fish feeding behavior, the speed of locomotor reactions in particular. It has been also demonstrated that 5-HT, as a rule, changes the level of carp intestinal mucosa proteolytic activity and especially amylolytic activity in comparison with that of the control fish during all experiment.

The level of glycemia decreases followed by sharp increase in 3 hours after 5-HT injection in comparison with that of the control fish. The pattern of changes in the relative glycemic level regarding the control group during first hour of observation is partly similar to that of enzyme activity. These facts evidence not only of the involvement of the brain's serotoninergic system in the regulation of the fish food intake but also of an important role of nutrition and, possibly, digestion under serotonin effects. In its turn, 5-HT can directly or indirectly influence the processes of digestion. However the mechanisms of 5-HT effect on digestive processes as well as feeding behavior and metabolism in fish need further investigation.

Acknowledgments

I am sincerely grateful to Dr. D. V. Garina and P. V. Rusanova for the participation in the experiments, as well as Dr.D. D. Pavlov for the linguistic help. This study was supported by the Russian Foundation for Basic Research, project 13-04-00248.

References

Abramets, I. I., Komissarov, I. V., Samoilovich, I. M., Shovtuta, V. I. Action of serotonin on smooth muscles. *Bull. Exp. Biol. Med.*, 1977 83, 174–177.

Anand, B. K., Brobeck, J. R. Localization of a "feeding center" in the hypothalamus of the rat. *Proc. Soc. Exp. Biol. Med.,* 1951 77, 323–324.

Anderson, C. Evidence for 5-HT-containing intrinsic neurons in teleost intestine. *Cell Tissue Res.* 1983 230(2), 377–386.

Anokhin, P. K. *Essays on the physiology of functional systems.* Meditsina, Moscow. Russia. 1975, 238 (In Russian).

Azmitia, E. C. Serotonin neurons, neuroplasticity, and homeostasis of neural tissue. *Neuropsychopharmacology*, 1999 21, 33S–45S.

Barton, B. A., Morgan, J. D., Vijayan, M. M. Physiological and condition-related indicators of environmental stress in fish. 2002. In: Adams, S. M. (Ed.). *Biological Indicators of Aquatic Ecosystem Stress.* American Fisheries Society, Bethesda, MD, 111–148.

Bernier, N. J., Peter, R. E. The hypothalamic-pituitary-interregnal axis and the control of food intake in teleost fish. *Comp. Biochem. Physiol.*, 2001 129 B, 639–644.

Barry, M. J. Effects of fluoxetine on the swimming and behavioural responses of the Arabian killifish. *Ecotoxicol.*, 2013 22, 425–432.

Bicego, K. C., Barros, R. C. H., Branco, L. G. S. Physiology of temperature regulation: comparative aspects. *Comp. Biochem. Physiol.*, 2007 147 A 616–639.

Buddington, R. K., Krogdahl, Å. Hormonal regulation of the fish gastrointestinal tract. *Comp. Biochem. Physiol.*, 2004 139 A, 261–271.

Caamaño-Tubío, R. I., Pérez, J., Ferreiro, S., Aldegunde, M. Peripheral serotonin dynamics in the rainbow trout (Oncorhynchus mykiss). *Comp. Biochem. Physiol.*, 2007 129 C, 245–255.

De Almeida, R. M. M., Ferrari, P. F., Parmigiani, S., Miczek, K. A. Escalated aggressive behavior: Dopamine, serotonin and GABA. *Europ. J. Pharmacol.*, 2005 526, 51–64.

De Pedro, N., Bjornsson, B. T. Regulation of food intake by neiro-peptides and hormones. In: *Food intake in fish*. Houlihan, D., Boujard, T., Jobling, M. (eds). Blackwell Sci. 2001 Ch. 12, 269–296.

De Pedro, N., Gansedo, B., Alonso-Gomez, A. L., Delgado, M. J., Alonso-Bedate, M. Role of corticotropin releasing factor (CRF) as a food intake regulator in goldfish. *Physiol. Behav.*, 1993 53, 517–520.

De Pedro, N., Pinillos, M. L., Valenciano, A. I., Alonso-Bedate, M., Delgado, M. J. Inhibitory effect of serotonin on feeding behavior in goldfish: Involvement of CRF. *Peptides*, 1998 19, 505–511.

Dinan, T. G. Serotonin and the regulation of hypothalamic-pituitary-adrenal axis function. *Life Sci.*, 1996 58, 1683–1694.

Donovan, M. H., Tecott, L. H. Serotonin and the regulation of mammalian energy balance. *Front Neurosci.*, 2013 7(36), 1–15.

Emerson, A. J., Kappenman, D. P., Ronan, P. J., Renner, K. J., Summers, C. H. Stress induces rapid changes in serotonergic activity: Restraint and exertion. *Behav. Brain Res.*, 2000 111, 83–92.

Essman, W. B., Serotonin distribution in tissues and fluids. In: Essman, W. B. (Ed.), *Serotonin in Health and Disease*. Availability, Localization and Disposition, SP Med. Sci. Books. 1978 I, 15–179.

Fontaine, Y. A. Hormones in Fishes. In: *Biochemical and biophysical perspective in marine biology*. London etc.: Acad. Press. 1975 2, 139–212.

Forster, M. E., Forster, A. H., Davison, W. Effects of serotonin, adrenaline and other vasoactive drugs on the branchial blood vessels of the Antarctic fish *Pagothenia borchgrevinki*. *Fish Physiol. Biochem.*, 1998 19, 103–109.

Fujimiya, M., Okumiya, K., Kuwahara, A. Immunoelectron microscopic study of the luminal release of serotonin from rat enterochromaffin cells induced by high intraluminal pressure. *Histochem. Cell Biol.*, 1997 108, 105–113.

Fuller, R. W. Serotonin receptors involved in regulation of pituitary-adrenocortical function in rats. *Behav. Brain Res.*, 1986 73, 215–219.

Gaworecki, K. M., Klaine, S. J. Behavior and biochemical responses of hybrid striped bass during and after fluoxetine exposure. *Aquat. Toxicol.*, 2008 88, 207–213.

Haider, S., Haleem, D. J. Decreases of brain serotonin following a food restriction schedule of 4 weeks in male and female rats. *Med. Sci. Monit.*, 2000 6, 1061–1067.

Hoebel, B. G. Neuroscience and appetitive behavior research: 25 years. *Appetite*, 1997 29 119–133.

Höglund, E., Bakke, M. J., Øverli, Ø., Winberg, S., Nilsson, G. E. Suppression of aggressive behaviour in juvenile Atlantic cod (*Gadus morhua*) by l-tryptophan supplementation. *Aquaculture*, 2005 249, 525–531.

Holmgren, S., Nilsson, S. Bombesin-, gastrin/CCK- 5-hydroxytryptamine-, neurotensin-, somatostatin-, and VIP-like immunoreactivity and catecholamine fluorescence in the gut of elasmobranch, *Squalus acanthias*. *Cell Tissue Res.*, 1983 234, 595–618.

Janvier, J.-J. Mediation of serotonin-induced branchial vasoconstriction by a cholinergic and muscarinic response *in vivo* in European eel *Anguilla anguilla*. *Fish Physiol. Biochem.*, 1997 16(1) 85–92.

Kah, O., Chambolle, P. Serotonin in the brain of the goldfish *Carassius auratus*. An immunocytochemical study. *Cell Tissue Res.*, 1983 234(2), 319–333.

Kapoor, B. C., Smit, H., Verighina, I. A. The alimentary canal and digestion in teleosts. *Advances in marine biology*. New York. 1975 13, 109–239.

Kaya, F., van Duin, C. T. M., Veenendaal, G. H., van Miert, A. S. Food intake and rumen motility in dwarf goats. Effects of some serotonin receptor agonists and antagonists. *Vet. Res. Commun.*, 1992 16(5) 379–390.

Kovacs, A., Harsing, Jr L. G., Szenasi, G. Vasoconstrictor 5-HT receptors in the smooth muscle of the rat middle cerebral artery. *Europ. J. Pharm.*, 2012 689(1-3), 160–164.

Kumar, P., Saurabh, Sh., Pal, A. K., Sahu, N. P., Arasu, A. R. T. Stress mitigating and growth enhancing effect of dietary tryptophan in rohu (*Labeo rohita*, Hamilton, 1822) fingerlings. *Fish Physiol. Biochem.*, 2014 40, 1325–133.

Kuz'mina, V. V. Hormonal Regulation of Metabolism and Processes of Exotrophy in Fish. Polyfunctionality and Polypotentiality (Review). *J. Evol. Biochem. Physiol.*, 2000 36(6) 670–684.

Kuz'mina, V. V. *Physiological and biochemical principles of exotrophy processes in fish.* Moscow: Nauka. 2005 300 (in Russian).

Kuz'mina, V. V. Influence of serotonin on the activity of digestive hydrolases in carp. *Problems. biol. product. animal.*, 2013 4, 53–60 (in Russian).

Kuz'mina, V. V., Garina, D. V. Glucose, insulin and adrenaline effects on some aspects of fish feeding behavior. *J. Evol. Biochem. Physiol.*, 2001 37 (2) 154–160.

Kuz'mina, V. V., Garina, D. V. The effects of peripherally injected serotonin on feeding and locomotor activities in carps *Cyprinus carpio* L. *Biol. Inland Waters*, 2013 6(1), 62–69.

Kuz'mina, V. V., Garina, D. V., Yablochkina, E. V. The influence of epinephrine on the process of polysaccharide hydrolysis, glycemic level and the concentration of glycogen in the tissues. *J. Evol. Biochem. Physiol.*, 2003 39(2), 140–146.

Kuz'mina, V. V., Ushakova, N. V., Rusanova, P. V. The effects of exogenous serotonin on feeding behaviour of carp *Cyprinus carpio* L. *Bull. Univer. Mordovia*, 2010 1 59–64(in Russian) .

Lam, D. D., Heisler, L. K. Serotonin and energy balance: molecular mechanisms and implications for type 2 diabetes. *Expert Rev. Mol. Med.*, 2007 9, 1–24.

Larson, E. T., Summers, C. H. Serotonin reverses dominant social status. *Behav. Brain Res.*, 2001 121, 95–102.

Le Feuvre, R. A., Aisenthal, L., Rothwell, N. J. Involvement of corticotrophin releasing factor (CRF) in the thermogenic and anorexic actions of serotonin (5-HT) and related compounds. *Brain Res.*, 1991 555, 245–250.

Leibowitz, S. F. Brain monoamines and peptides: role in the control of eating behaviour. *Fed. Proc.*, 1986 45, 1396–1403.

Leibowitz, S. F., Alexander, J. T. Hypothalamic serotonin in control of eating behavior, meal size, and body weight. *Biol. Psychiatry.*, 1998 144 (9), 851–864.

Leonard, B. E. The HPA and immune axes in stress: the involvement of the serotonergic system. *Europ. Psychiatry*, 2005 20, S302–S306.

Lepage, O., Larson, E. T., Mayer, I., Winberg, S. Serotonin, but not melatonin, plays a role in shaping dominant–subordinate relationships and aggression in rainbow trout. *Hormon. Behav.*, 2005 48, 233–242.

Lin, X., Volkoff, H., Narnaware, Y., Bernier, J., Peyon, P., Peter, R. E. Brain regulation of feeding behavior and food intake in fish. *Compar. Biochem. Physiol.* 2000 126 A, 415–434.

Lillesaar, C. The serotonergic system in fish. *J. Chem. Neuroanat.* 2011 41(4), 294–308.

Lowry, Ch. A., Moore, F. L. Regulation of behavioral responses by corticotrophin-releasing factor. *Gen. Compar. Endocrinol.*, 2006 146, 19–27.

Lyons, P. M., Truswell, A. S. Serotonin precursor influenced by type of carbohydrate meal in healthy adults. *Am. J. Clin. Nutr.* 1988 47, 433–439.

MacKenzie, D. S., VanPutte, C. M., Leiner, K. A. Nutrient regulation of endocrine function in fish. *Aquaculture* 1998 161, 3–25.

Marston, O. J., Garfield, A. S., Heisler, L. K. Role of central serotonin and melanocortin systems in the control of energy balance. *Eur. J. Pharmacol.* 2011 660, 70–79.

Moran, T. H., Kinzig, K. P. Gastrointestinal satiety signals. II. Cholecystokinin. *Am. J. Physiol. Gastrointest. Liver Physiol.* 2004 286, G183–G188.

Nelson, L. E., Sheridan, M. A. Gastroenteropancreatic hormones and metabolism in fish. *Gen. Compar. Endocrin.* 2006 148, 116–124.

Nosdrachev, A. D. Neurophysiology of the ganglia of intramural plexus of the digestive tract. In: Bacladjan, O. G., Bakuntz, F. A., Veller, M. N., Buligin, I. A., Kibiakov, A. V., et al. (eds). Physiology of the autonomic nervous system. *Handbook of physiology.* Leningrad: Nauka, 1981 152–186 (In Russian).

Overli, O., Winberg, S., Damsgard, B., Jobling, M. Food intake and spontaneous swimming activity in Arctic char (*Salvelinus alpinus*): role of brain serotonergic activity and social interactions. *Can. J. Zool.*, 1998 76 1366–1370.

Pavlov, I. P. About the feeding center. In: *The complete works.* Moscow-Leningrad. 1951 3(1), 147-158 (in Russian).

Petrov, T., Krukoff, T. L., Jhamandas, J. H. The hypothalamic paraventricular and lateral parabrachial nuclei receive collaterals from raphe nucleus neurons: a combined double retrograde and immunocytochemical study. *J. Comp. Neurol.* 1992 318, 18–26.

Piomelli, D., Tota, B. Different distribution of serotonin in an elasmobranch (*Scyliorhinus stellaris*) and in a teleost (*Conger conger*) fish. *Comp. Biochem. Physiol.*, 1983. 74C, 139–142.

Plisetskaya, E. M. *Hormonal regulation of carbohydrate metabolism in lower vertebrates.* Leningrad: Nauka. 1975, 215.

Plisetskaya, E. M., Buchelli-Narvaez, L. I., Hardy, R. W., Dickhoff, W. W. Effects of injected and dietary arginine on plasma insulin levels and growth of Pacific salmon and rainbow trout. *Comp. Biochem. Physiol.* 1991 A 98, 165–170.

Rothwell, N. J., Stock, M. J. Effect of diet and fenfluramine on thermogenesis in the rat: possible involvement of serotonergic mechanisms. *Int. J. Obes.* 1987 11, 319–324.

Rubio, V. C., Sanchez-Vazquez, F. J., Madrid, J. A. Oral serotonin administration affects the quantity and the quality of macronutrients selection in European see bass *Dicentrarchus labrax* L. *Physiol. Behavior.*, 2006 87, 7–15.

Ritche, T. C., Livingston, C. A., Hughes, M. G., McAdoo, D. J., Leonard, R. B. The distribution of serotonin in the CNS of an elasmobranch fish: immunocytochemical and biochemical studies in the Atlantic stringray *Dasyatus sabina. J. Comp. Neurol.* 1983 221 (4), 429-443.

Ruchin, A. B. Peculiarities of growth and energy metabolism in carps (*Cyprinus carpio*) in conditions of different light regimes. *Zool. J.*, 2001 80(4), 433–437(in Russian).

Ruibal, C., Soengas, J. L., Aldegunde, M. Brain serotonin and the control of food intake in rainbow trout (*Oncorhynchus mykiss*): Effects of changes in plasma glucose levels. *J. Comp. Physiol.*, 2002 188 A, 479–484.

Schparkovsky, I. A. *Physiology of fish digestion: locomotor function.* Leningrad: Nauka, 1986 176 (In Russian).

Schuba, M. F. Mechanisms of the action of mediators on the smooth muscle cells. In: Bacladjan, O. G., Bakuntz, F. A., Veller, M. N., Buligin, I. A., Kibiakov, A. V., et al. (eds). *Physiology of the autonomic nervous system. Handbook of physiology.* Leningrad: Nauka, 1981 105–128 (In Russian).

Schwartz, D. H., McClane, S., Hernandez, L., Hoebel, B. G. Feeding increases extracellular serotonin in the lateral hypothalamus of the rat as measured by microdialysis. *Brain Res.* 1989 479, 349–354.

Sebert, P., Barthelemy, L., Caroff, J. Serotonin levels in fish brain: effects of hydrostatic pressure and water temperature. *Experientia* 1985 41, 1429–1430.

Silva, P. I. M., Martins, C. I. M., Höglund, E., Gjøen, H. M., Øverli, Ø. Feeding motivation as a personality trait in Nile tilapia (*Oreochromis niloticus*): role of serotonergic neurotransmission. *Fish Physiol. Biochem.* 2014 40 5 1547–1557.

Simansky, K. J. Serotonengic control of the organization of feeding and satiety. *Behav. Brain Res.* 1996 73 37–42.

Soengas, J. L., Aldegunde, M. Energy metabolism of fish brain. *Compar. Biochem. Physiol.* 2002 131B 271–296.

Stoddard, P. K., Markham, M. R., Salazar, V. L. Serotonin modulates the electric wave form of the gymnotiform electric fish *Brachyhypopomus pinnicaudatus*. *J. Exp. Biol.* 2008 206, 1353–1362.

Terenina, N. B., Gustafsson, M. K. S. *Neurotransmitters in helminths (biogenic amines, nitric oxide)* Moscow: Nauka, 2003 179 pp (in Russian).

Tsai, C.-L., Wang, L.-H., Tsai, C.-C. Role of Serotonin, c-Aminobutyric Acid, and Glutamate in the Behavioral Thermoregulation of Female Tilapia during the Prespawning Phase. *J. Exp. Zool.* 2002 293, 443–449.

Volkoff, H., Canosa, L. F., Unniappan, S., Cerdá-Reverter, J. M., Bernier, N. J., Kelly, S. P., Peter, R. E., Neuropeptides and the control of food intake in fish. *Gen. Comp. Endocrinol.* 2005 142, 3–19.

Watanabe, H., Akasaka, D., Ogasawara, H., Sato, K., et al. Peripheral serotonin enhances lipid metabolism by accelerating bile acid turnover. *Endocrinology* 2010a 151, 4776–4786.

Watanabe, R. L., Andrade, I. S., Telles, M. M., Albuquerque, K. T., Nascimento, C. M., Oyama, L. M., Casarini, D. E., Ribeiro, E. B. Long-term consumption of fish oil-enriched diet impairs serotonin hypophagia in rats. *Cell Mol. Neurobiol.* 2010b 30(7), 1025–1033.

Winberg, S., Carter, C. G., McCarthy, I., He, Z.-Y., Nilsson, G. E., Houlihan, D. F. Feeding rank and brain serotonergic activity in rainbow trout *Oncorhynchus mykiss*. *J. Exp. Biol.* 1993a 179, 197–211.

Winberg, S., Nilsson, G. E., Spruijt, B. E., Hoglund, U. Spontaneous locomotor activity in Arctic charr measured by a computerized imaging technique: role of brain serotonergic activity. *J. Exp. Biol.* 1993b, 179 213–232.

Winberg, S., Nilsson, A., Hylland, P., Söderstöm, V., Nilsson, G. E. Serotonin as a regulator of hypothalamic-pituitary-interrenal activity in teleost fish. *Neuroscience Letters* 1997 230, 113–116.

Yamada, J., Sugimoto, Y., Kimura, I., Takeuchi, N., Horisaka, K. Serotonin-induced hypoglycemia and increased serum insulin levels in mice. *Life Sci.* 1989 45, 1931–1936.

Yamada, J., Sugimoto, Y., Ujikawa, M. Involvement of leptin in hypophagia induced by the serotonin precursor 5-hydroxytryptophan (5-HTP) in mice. *Biol. Pharm. Bull.* 2006 29, 557–559.

Zubizarreta, L., Perrone, R., Stoddard, P. K., Costa, G., Silva, A. C. Differential serotonergic modulation of two types of aggression in weakly electric fish. *Front. Behav. Neurosci.* 2012 6, 77–81.

In: New Developments in Serotonin Research ISBN: 978-1-63482-220-6
Editor: Ming D. Li

Chapter 6

Engagement of Serotonin-Modulating Anticonsolidation Protein in Memory Formation and Suppression of Drug Addiction and Epileptic Seizures

Arif A. Mekhtiev[*], *Elmira N. Panahova,*
Afag M. Rashidova, and Shovqi B. Guseinov
Abdulla Karaev Institute of Physiology, Baku, AZ, Azerbaijan

Abstract

We conducted experiments involving application of a novel serotonin-modulating anticonsolidation protein (SMAP), which was identified originally in the rat brain cortex and purified from the whole brain. It has an M_r of 186 kDa and shows a linear relation with the serotonin concentration (Mekhtiev, 2000; Mekhtiev et al., 2004). We describe the effects of SMAP on memory formation in Wistar rats, which was examined in conditioned passive avoidance and shuttle box models with electric shock reinforcement and the instrumental model of alternative choice with food reinforcement (Guseinov, Mekhtiev, 2013)

[*] Corresponding Author: Arif A. Mekhtiev, PhD, Department of Physiology, 2 Sharif-zadeh St., AZ1100 Azerbaijan. E-mail: arifmekht@yahoo.com.

and the process of reconsolidation. In the morphine self-administration model, the SMAP concentration in the cingulate cortex of addicted rats increased more than 67%, whereas there was no change in the concentration in the hypothalamus. The effects of a single intramuscular (IM) dose of anti-SMAP polyclonal antibodies in the drug-addicted animals were analyzed (Mekhtiev et al., 2014). In experiments with the *Shinshilla* rabbit, the effects of IM administration of SMAP on penicillin-induced epileptic seizures and of microinjection of polyclonal anti-SMAP antibodies into the amygdala on the development of epileptic seizures were studied.

Keywords: drug addiction, epileptic seizures, memory consolidation, serotonin-modulating anticonsolidation protein

Introduction

As the effects of serotonin on nerve cells are mediated by activation of specific genes, and the synthesis of relevant proteins is regulated by serotonin (Barziali et al., 1989; Abel, Kandel, 1998), the involvement of the serotonergic system in memory formation and other serotonin-related processes requires investigation. A novel serotonin-modulating anticonsolidation protein (SMAP) was identified in rat brain cortex using electrophoresis in continuous gradients (4–30%) of polyacrylamide gel (Mekhtiev, 2000). The SMAP concentration was found to be upregulated by more than 84% ($p < 0.01$) 24 h after administration of anti-SMP-69 protein polyclonal antibodies into the lateral ventricle of the rat brain. The SMP-69 protein had earlier been identified after injection of 10^{-4} M serotonin solution into the occipital cortex of narcotized Wistar rats for 30 min. Thereafter, it was purified from the whole rat brain and found to have an M_r of 69 kDa and a p*I* of 6.0 and to increase in concentration as serotonin decreased (Mekhtiev et al., 1996). By electrophoresis, SMAP was a heavy-weight molecule with low mobility, having an M_r of 186 kDa and bearing fraction number of 28. Then SMAP was purified from whole rat brains under screening control of an ELISA with γ-globulins. The main steps of SMAP purification were partial precipitation of water-soluble proteins in the range of a 0–40% concentration of ammonium sulfate and gel chromatography on Sephadex G-150 columns (3 × 60 cm). To measure the SMAP concentration in the animal's tissues with the ELISA, anti-SMAP polyclonal γ-globulins were produced through 5–6-month immunization of rabbits with purified SMAP (300 μg per injection), always in combination with an equal amount of complete Freund's adjuvant (Mekhtiev, 2000).

All experiments undertaken on animals were conducted according to the regulations of the Directive of the Council of the European Society (86/609/EEC).

1. Participation of SMAP in Memory Formation

The studies of SMAP effects on memory processes were conducted using the conditioned passive avoidance model in male Wistar rats. SMAP was administered in a dose of 10 μL (concentration 1.5 mg/mL) into the lateral brain ventricle 24 h prior to the learning sessions. Such treatment led to impairment of memory, as judged by a number of animals' entering into the dark chamber. That is, in the re-test sessions, 87% of the animals in the experimental group entered the dark chamber, whereas only 8% of animals in the control group did so (p <0.001; Student's *t*-criterion) (Figure 1). At the same time, SMAP administration 48 h after completion of learning sessions did not have a significant effect on memory, indicating that SMAP impairs memory consolidation but does not have any effect on storage or retrieval of memory traces (Mekhtiev, 2000). This activity was the basis of our naming this protein serotonin-modulating anticonsolidation protein.

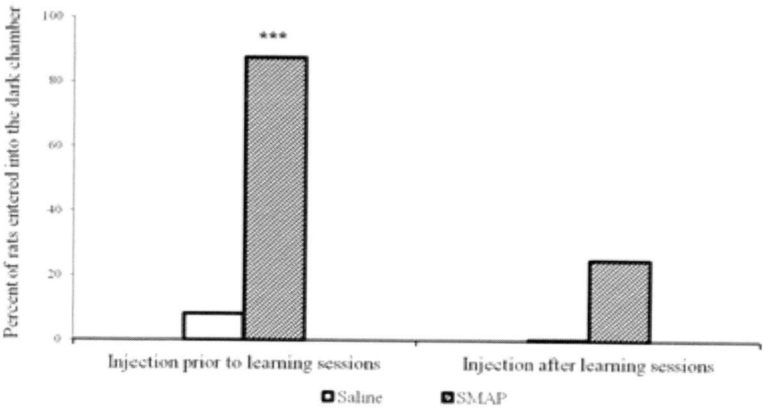

Figure 1. Effect of intraventricular SMAP administration on rats entering the dark chamber during retest sessions in the passive avoidance model (n = 16). ***p <0.001.

In the next studies, the role of SMAP in memory formation was analyzed using the many-time shuttle box model (Guseinov, Mekhtiev, 2013). In the first series, the animals were assigned to three groups: (1) intact animals; (2) active control group; (3) experimental group. Intact animals were kept in standard cages. Animals of the active control group were placed in the shuttle box and presented with non-paired sounds (6 sec) from each side alternately and received painful electrical stimuli of 0.8 mA (6 sec) via the grid floor of the cage. The number of stimuli presented was identical to the number of stimuli given to the animals of the corresponding experimental group. Those animals were presented with the sound (6 sec) on the right or left side, followed without delay by delivery of a painful electrical stimulus (6 sec) in the same part of the chamber. The animals were trained up to the 50% or 80% learning criterion, calculated as the ratio of the number of correctly performed trials to the total number of trials. On achieving the learning criterion, animals were sacrificed; the occipital and parietal areas of the cortex were taken from both hemispheres, frozen, and stored at –70°C until use.

The SMAP concentrations in the rat cerebral cortex were measured by indirect ELISA in the plates with intermediate adsorption (Sigma, USA) using rabbit polyclonal anti-SMAP immunoglobulins. The animals of the experimental group trained to the 80% criterion showed downregulation of SMAP in the parietal ($p < 0.05$) and occipital ($p < 0.05$; Student's *t*-criterion) cortex relative to the animals of the active control group receiving the same amount of acoustic and electric stimuli (Figure 2; Guseinov, Mekhtiev, 2013).

In the second series of studies, cannulas were inserted into the lateral ventricles of the brains of anesthetized rats of the experimental and control groups. Three days later, the animals were started on administration of SMAP (10 μL of a 1.5 mg/mL solution) or inactivated SMAP (heated in a water bath at 55°C for 30 min) daily for several days. Forty minutes after each injection, the animals were subjected to the training procedures up to the 50% learning criterion. There were no apparent differences in the rate of memory formation in the native and control groups, and these groups reached the 50% criterion for 4 days. After presentation of the acoustic stimulus alone, the animals passed to another chamber of the box. In these experiments, the behavior of the experimental animals was quite different: in the first 2 days, these animals did not show any sign of memory formation. Although they reached the 50% learning criterion by 7 days ($p < 0.001$; Student's *t*-criterion) (Figure 3), they did not acquire stable memory (Guseinov, Mekhtiev, 2013).

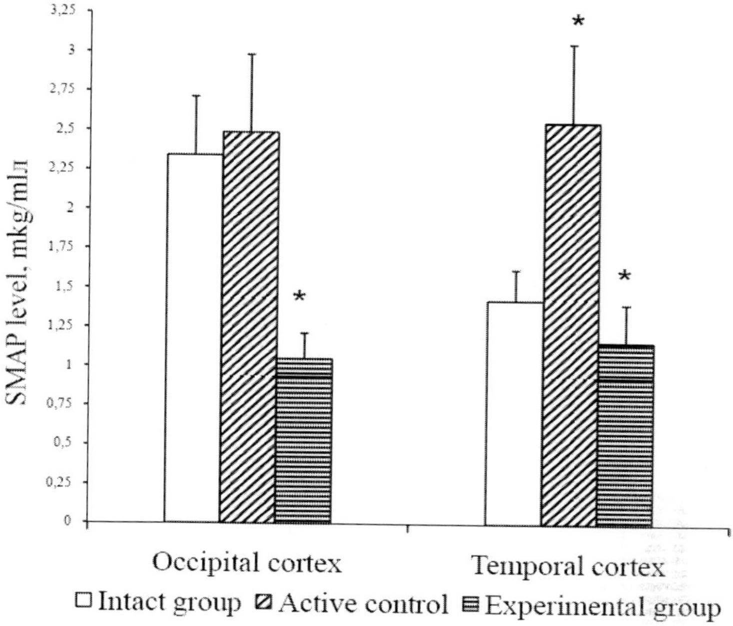

Figure 2. Changes in SMAP concentration in the brain cortex of rats achieving 80% of learning criterion in the shuttle box model (n = 10). *p <0.05.

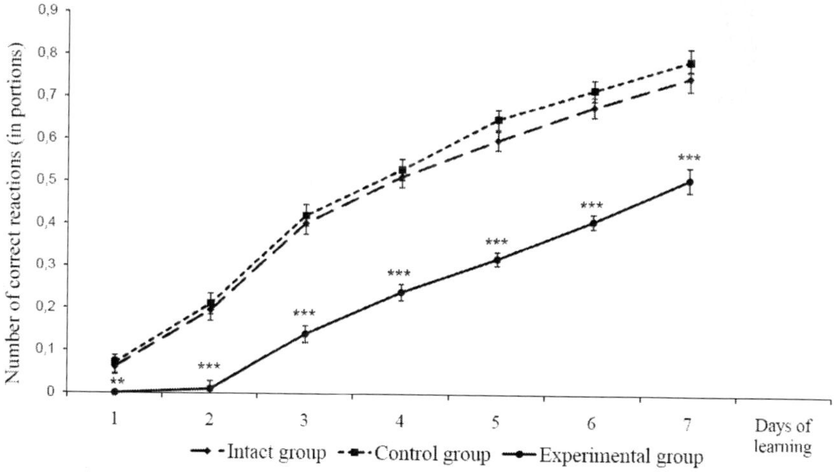

Figure 3. Dynamics of formation of task mastery in rats in shuttle box for 7 days of learning sessions (n = 12). **p <0.01; ***p <0.001 by Student's *t*-criterion.

In the next study, involvement of SMAP in memory processes was analyzed in the model of alternative running with food reinforcement. The experimental box had dimensions of 60 × 60 cm, and two platforms of 20 × 10 cm were fixed on the left and right walls at a height of 16 cm from the floor. Food-deprived animals were trained to run to the left and right platforms in turn and to get food pellets by climbing on them. This study included two series similar to those of the study of SMAP role in memory processes in the shuttle box. In this model, as a result of memory formation, significant downregulation of SMAP was noticed in the parietal cortex of the rats that achieved the 80% learning criterion ($p < 0.001$; Student's *t*-criterion) (Figure 4).

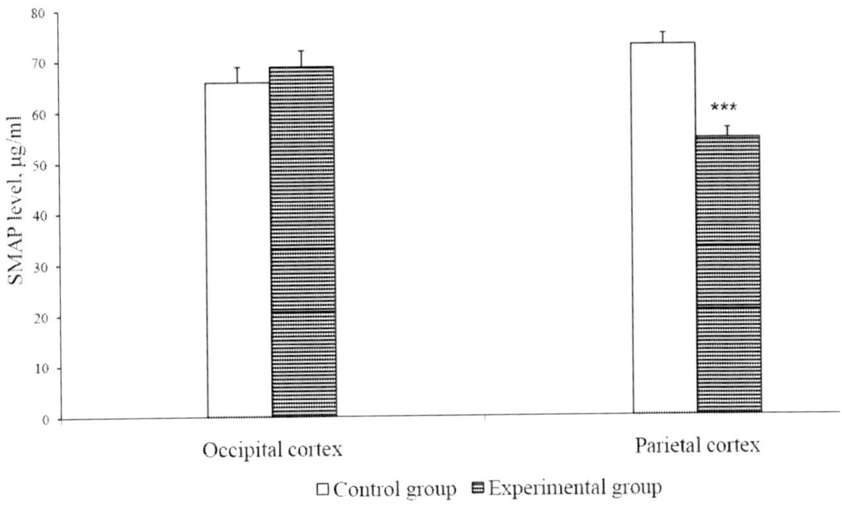

Figure 4. Change in SMAP concentration in the brain cortex in rats trained in the model of alternative running ($n = 12$). ***$p < 0.001$ by Student's *t*-criterion.

Next, SMAP was injected into the lateral ventricle, which led to total impairment of memory consolidation, whereas intact and control animals (given heat-inactivated SMAP) mastered this task and reached the 50% learning criterion with 6 days of daily training ($p < 0.001$; Student's *t*-criterion) (Figure 5).

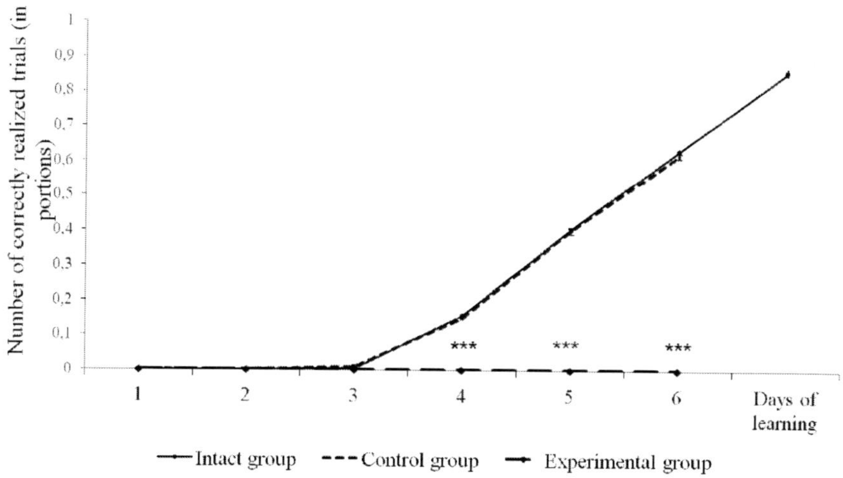

Figure 5. Dynamics of formation of task of alternative running in rats (n = 12). ***p <0.001 by Student's *t*-criterion.

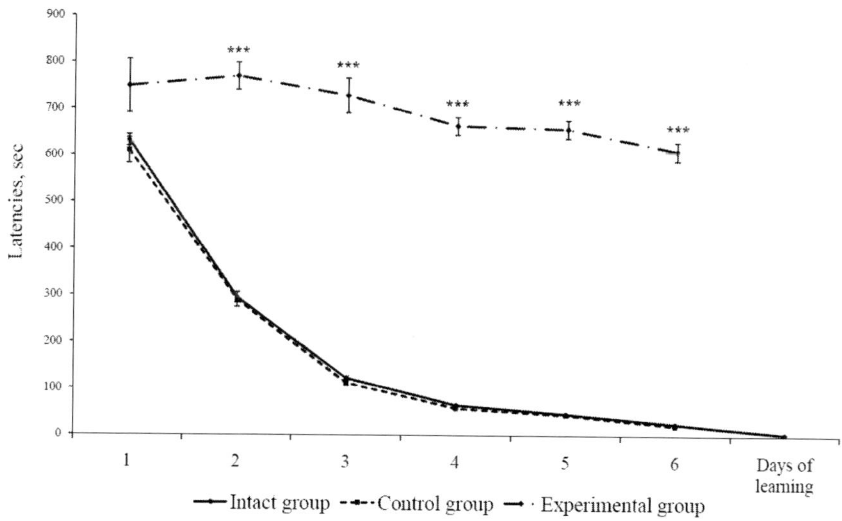

Figure 6. Dynamics of latencies of the first running to the platform in the alternative running model on different days of learning (n = 10). ***p <0.001 by Student's *t*-criterion.

Values of latencies of the first successful trial in the intact and control animals showed a constant decrease over the course of learning, whereas the latencies in the experimental animals remained unchanged ($p < 0.001$; Student's *t*-criterion) (Figure 6).

Throughout course of several days of learning, the animals of the experimental group always behaved as if they were being put in the experimental box for the first time. They did not become familiar with the surrounding signals and demonstrated high fear and anxiety. In a separate study, we analyzed the effects of SMAP on the degree of food motivation, and intracerebral SMAP administration did not decrease it. Thus, impairment of memory processes in the experimental group could not be explained by SMAP lowering of the extent of food motivation (Guseinov, Mekhtiev, *in press*).

In another series of studies, we investigated the involvement of SMAP in memory reconsolidation. Intact male Wistar rats were trained in the shuttle box model until they achieved the 80% criterion for 4 days. Then they were put back in their home cages for 15 days, after which, they were taken out to study reconsolidation. For this purpose, SMAP was injected into the lateral ventricle, and 40 min later (as in the memory studies), the animals were put into the experimental box and subjected to the acoustic stimulus. There were no signs of memory impairment (Figure 7), as usually occurs in reconsolidation studies when the conditioned stimulus is presented to the animals having memory traces after intracerebral administration of inhibitors of protein synthesis or agonists of neurotransmitter receptors (Kemenes et al., 2006; Nikitin, Solntseva, 2010).

The ineffectiveness of SMAP in impairing memory traces in the reconsolidation studies probably means that different (other than SMAP) protein(s) underlies reconsolidation. This conjecture should be analyzed thoroughly in different behavioral models.

In sum, SMAP possesses anticonsolidation activity in rat models with negative and positive reinforcement if it is injected intracerebrally prior to the learning sessions.

Moreover, it is downregulated after formation of memory traces in the brain cortex. The SMAP concentration in the brain cortex does not change during memory reconsolidation, strongly suggesting that different proteins underlie memory formation and reconsolidation.

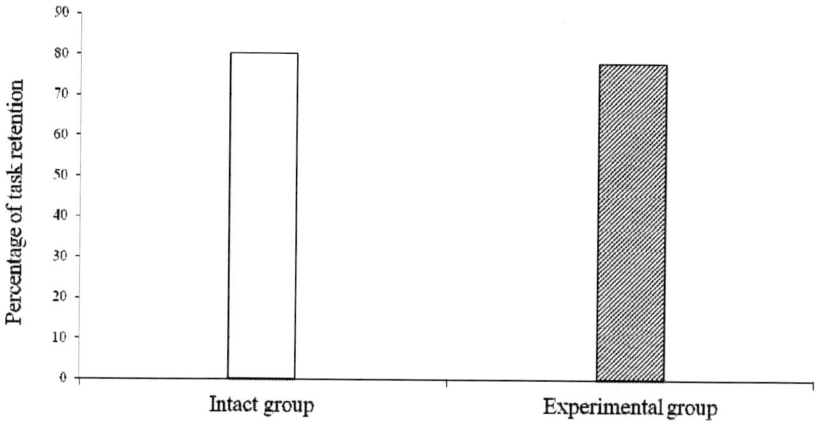

Figure 7. Effects of SMAP on memory reconsolidation in the shuttle box model (n = 10).

2. Engagement of Smap in Drug Addiction

The serotonergic system is significantly involved in the creation of drug addiction. After administration of addicting drugs, an acute increase in serotonin is observed, and its concentration normalizes gradually as the drug is metabolized and excreted. In the *in vivo* application of the microdialysis method in rats, after cocaine self-administration, an increase of more than 340% in the baseline extracellular serotonin concentration persists for 12 h in the nucleus accumbens (Parsons et al., 1995). When there is regular administration of addicting drugs, the serotonin content in the whole organism decreases if there is sharply a pause in drug administration, which induces grave suffering in patients (the abstinence syndrome). In the same work, withdrawing rats from cocaine resulted in a significant decrease in the serotonin concentration in the nucleus accumbens relative to the controls (0.6 ± 0.3 nM vs. 2.0 ± 0.5 nM, respectively) (Parsons et al, 1995).

An increase in the serotonin concentration during drug administration for addiction treatment does not simply coincide in time with dependency establishment, but underlies it. Direct microapplication of serotonin to hypothalamic neurons, which are involved in elaboration of drug dependency (revealed by intracellular registration of neuronal activity), leads to the same pattern of changes in neuronal activity observed after a single dose of morphine (Gasanov, Kuliyev, 1992). These data confirmed the specific role of serotonin as a neurotransmitter responsible for specific effects of addiction treatments on brain neurons.

Taking these finding together, we decided next to study active engagement of SMAP as a serotonin-modulating protein in establishment of drug addiction. For this purpose, addiction was created in Wistar rats using the morphine self-administration model. First, special catheters were introduced into the jugular veins of anesthetized (sodium ethaminal, 40 mg/kg) rats from the experimental group, and the thick ends of the catheters were fixed under the skin on the back of the neck. After recovery, the rats were put into the experimental box individually to acquire addiction to morphine. The experimental box (60 cm × 60 cm) contained two levers fixed to the back wall. One of them was connected to a peristaltic pump by silicone tubing. Each press of the lever by the rats resulted in automatic supply of 50 μL of 0.2% (100 μg) morphine into the vein.

In the first series of studies, after the animals became addicted to morphine and reached a standard dose, the animals were sacrificed, and the hypothalamus and cingulate cortex were removed. Water-soluble proteins were extracted for use as antigens in the indirect ELISA test. Rabbit anti-SMAP IgG, produced by 3–4-month immunization, was used as the first antibodies. In the rats with established drug addiction, SMAP was found to be upregulated by more than 67% in the cingulate cortex ($p = 0.01$; Mann-Whitney's U criterion), whereas its concentration did not change in the hypothalamus (Figure 8; Mekhtiev et al., 2014).

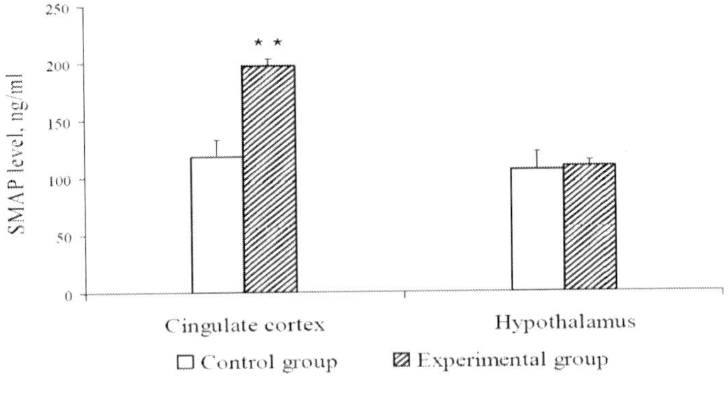

Figure 8. Changes in SMAP concentration in the brain of morphine-addicted animals ($n = 9$). $^{**}p < 0.01$.

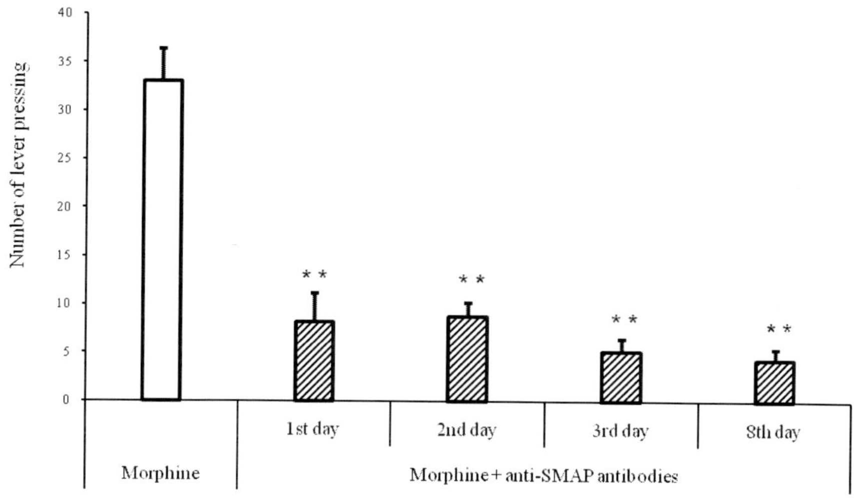

Figure 9. Effects of anti-SMAP antibodies on morphine intake in rats allowed morphine self-administration (n = 9). **p <0.008 by Fridman's χ^2 criterion.

In the second series of studies, after the rats reached constant morphine intake, they were assigned to two groups: (1) the experimental group, which received intraperitoneal rabbit anti-SMAP polyclonal antibodies (purified by immune affinity chromatography from rabbit anti-SMAP IgG on a CNBr-Sepharose column coupled with immobilized SMAP); and (2) the control group, which were given non-immune rabbit IgG. Both preparations were used at a ratio of 1 mg/100 g of body mass.

One injection of anti-SMAP antibodies to the addicted animals led to many-fold downregulation of the effect of morphine (p <0.008; Fridman's χ^2 criterion) (Figure 9)(Mekhtiev et al., 2014), which was detectable on the second, third, and eighth days after the injection (measurement of morphine intake for a longer time was difficult for technical reasons). When such animals were placed in the experimental box, they came up close to the lever, smelled it, but moved aside. However, control animals having constant extents of morphine intake did not change their consumption after administration of non-immune IgG (Figure 10). As SMAP upregulation was observed in the cingulate cortex of the addicted animals, its involvement in the formation and persistence of drug addiction is clear. Owing to the serotonin-modulating nature of SMAP and in view of serotonin's activity, one-time blockade of its activity with polyclonal antibodies, as in our studies, results in disruption of intracellular transduction of the serotonin signal, eliminating the positive reinforcement properties of morphine in addicted rats.

Thus, a single dose of anti-SMAP antibodies suppressed the positive reinforcement effect of morphine in drug-dependent animals. As mentioned above, we also found significant downregulation of SMAP in the cortex of animals trained in different models. Taken together, these results suggest that the prolonged effects of anti-SMAP antibodies might be related to induction of memory trace formation and the impossibility of expression of positive emotions based on artificial inactivation of SMAP (downregulation).

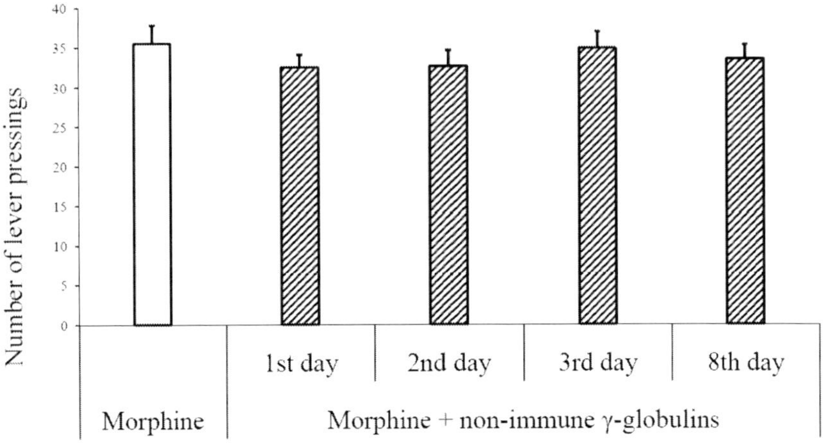

Figure 10. Effects of non-immune γ-globulins on morphine intake in rats allowed morphine self-administration (n = 10).

3. Role of Smap in the Action and Suppression of Epileptic Seizures

Numerous investigations have shown engagement of the brain serotonergic system in the pathogenesis of epilepsy. As has been proved by numerous studies in both animal models and humans, epilepsy develops against a background of low brain serotonin (Jobe, 2003), and serotonergic system activation exerts an anticonvulsant effect. In particular, serotonin 5-HT_{1A} receptor activation elicits a membrane hyperpolarizing response related to increased potassium conductance (Okuhara and Beck, 1994) and has an anticonvulsant effect *in vivo* as well as *in vitro*, including in hippocampal-kindled seizures in cats, intrahippocampal kainic acid-induced seizures in freely moving rats, and picrotoxin-, bicuculline-, and kainic acid-induced seizures in rat hippocampal slices.

Selective serotonin reuptake inhibitors (SSRIs), such as fluoxetine, possess prominent anti-seizure effects in several models (Hernandez et al., 2002). Generally, these agents elevate extracellular serotonin concentrations and inhibit both limbic and generalized seizures. Conversely, depletion of brain serotonin reduces the threshold for audiogenic, chemical, and electrically evoked convulsions.

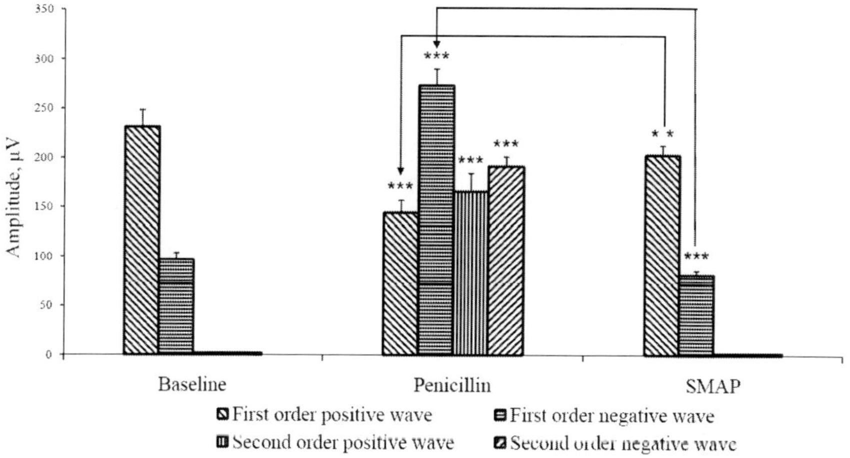

Figure 11. Changes of amplitudes of evoked potentials in the colliculus superior under penicillin microapplication into the amigdala and IM injection of SMAP in rabbits. *p <0.05; ***p <0.001.

The goal of the present study was to examine the role of SMAP in the development of epileptic activity in the rabbits. The studies were conducted on male *Shinshilla* rabbits. Epileptic seizures appeared 20–30 min after microapplication of a solution of penicillin sodium into the amygdala (10 μL; 300 units). During epileptic seizures, significant changes in the patterns of evoked potentials were registered in the superior colliculus (SC) and visual cortex (VC) in response to light flashes. In the SC, the amplitude of first-order positive waves decreased by more than 40% (p <0.001), and first-order negative wave amplitude was upregulated by more than 175% (p <0.001). Second-order positive and negative waves appeared (Figure 11). In contrast, in the VC, after the start of epileptic upregulation of the amplitude of first-order positive waves by more than 134% ($p < 0.001$), downregulation of the amplitude of first-order negative waves by more than 49% ($p < 0.001$) was noticed. The amplitude of second-order positive waves was upregulated by more than 90% ($p < 0.001$; Figure 12). Changes in the interictal spikes were observed in both structures. Prolonged clonic seizures occurred in the torso and limb muscles. In the control animals, seizures continued for more than 3 h, as judged by both external manifestations and electrographic indexes.

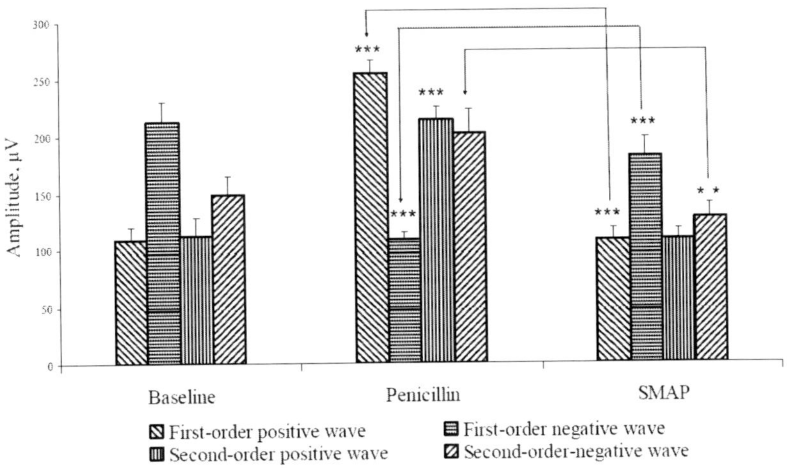

Figure 12. Changes of amplitudes of evoked potentials in visual cortex of rabbits after penicillin microapplication into the amydala and IM injection of SMAP. **p <0.01; ***p <0.001.

After the development of epileptic seizures, SMAP was administered IM (3.5 mL; 1.5 mg/mL) in buffered saline to the rabbits of the experimental group. To exclude non-specific effects of heterologous protein, the control animals were injected with inactivated SMAP (heating in water bath at 55°C for 30 min). After 30–40 min, inhibition of clonic seizures was registered. After 60 min of SMAP administration, interictal spikes were eliminated, and patterns of evoked potentials recovered in the opposite order: first in the CS, then in the VC. Particularly, in the CS, the amplitude of first-order positive waves became elevated, whereas the amplitude of the first-order negative waves diminished, becoming 87% ($p < 0.01$) and 84% ($p < 0.001$) of the original level, respectively (Figure 11). Under the effect of SMAP in the VC, restoration of the amplitudes of all waves relative to the original was noticed: first-order positive waves up at least 100%, first-order negative waves up 86%

or more (p <0.001), second-order positive waves up 97% (p <0.001), and second-order negative waves up 86% (p <0.01, Figure 12). In the control animals, IM administration of heat-inactivated SMAP did not have any effect on seizures and did not induce any electrographic indexes of epileptic activity.

In additional experiments, the permeability of the blood–brain barrier (BBB) to SMAP was analyzed. With application of the indirect ELISA and anti-SMAP IgG, it was shown that IM administration of SMAP in the same amount used in the first series of experiments led to a four-fold upregulation of SMAP in the liquor relative to its baseline concentration in the same animals after 50 min (56 ng/mL and 12.6 ng/mL, respectively). These data indicate a high degree of permeability of the BBB to SMAP.

In the second series of studies, microapplication of rabbit anti-SMAP polyclonal antibodies (10 μL, 1.5 mg/mL) to the amygdala brought about epileptic seizures after a delay of 2.5–3 h. In this case, both external manifestations, in the form of gritting of teeth, turn-by-turn contractions of mimic muscles, movements of the muscles of the torso and limbs, and characteristic changes in the patterns of evoked potentials, were noticed. Particularly in the CS, in response to light flashes, upregulation of the amplitude of first-order positive waves by more than 130% (p <0.001) and lack of changes in the amplitude of first-order negative waves was registered (Figure 13), whereas in the VC, the opposite patterns of evoked potentials were revealed: upregulation of the amplitude of first-order negative waves by more than 118%, lack of changes in the amplitude of first-order positive waves, and advent of second-order positive and negative waves (Figure 14). Administration of rabbit non-immune IgG into the amygdala of the control animals did not induce any changes in the amplitudes of evoked potentials in the CS or VC.

In the third series of studies, rabbits were immunized with SMAP (300 μg per immunization, always in combination with Freund's complete adjuvant) to produce anti-SMAP antibodies. Blood sampling of rabbits 10 days after the last injection showed high titers of anti-SMAP antibodies. Prior to immunization of rabbits with SMAP, maximum doses of penicillin solution (200 U) did not cause epileptic seizures in naïve animals. After 30 min of microapplication of 200 units of penicillin solution into the amygdala, external manifestations of epileptic seizures were observed for 4–5 min: clonic seizures, trembling of the whole body, frequent breathing, and unwilling urination. While fixing evoked potentials in the immunized rabbits after microapplication of subthreshold amounts of penicillin solution in the CS, upregulation of amplitudes of first-order positive and negative waves by more than 148% (p <0.05) and 100% (p <0.01), respectively, were seem, as well as spike activity (Figure 15). In the VC, upregulation of the amplitude of first-order waves by more than 79% (p <0.05), as well as upregulation of the amplitudes of second-order positive and negative waves, was observed (Figure 16). Spike activities were registered in the loci of first-order and second-order waves.

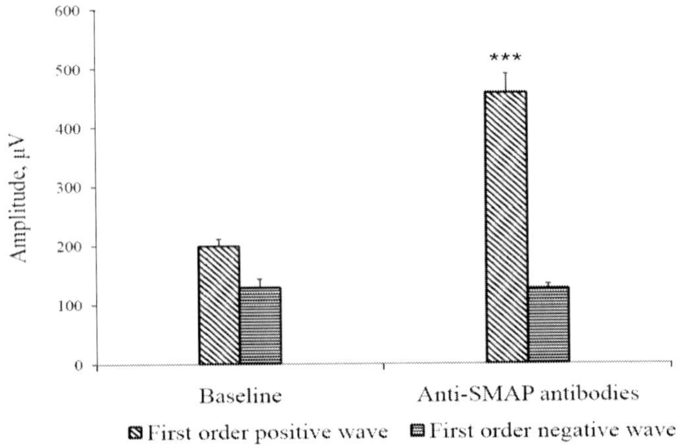

Figure 13. Changes of amplitudes of evoked potentials in the colliculum superior of rabbits given anti-SMAP antibodies. ***p <0.001 by Student's *t*-criterion.

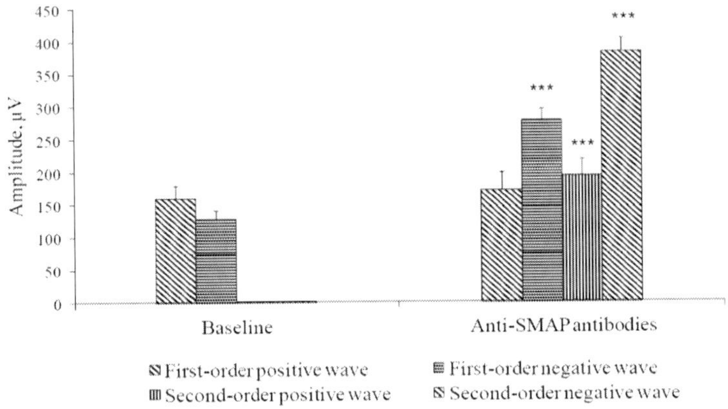

Figure 14. Changes of amplitudes of evoked potentials in the visual cortex of rabbits given anti-SMAP antibodies. ***p <0.001.

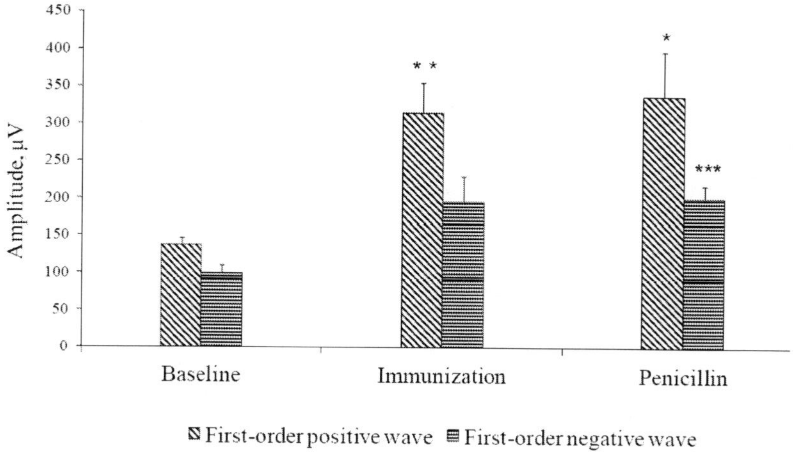

Figure 15. Changes of evoked potentials in the colliculum superior of rabbits receiving immunization with SMAP. *p <0.05; **p <0.01; **p <0.001.

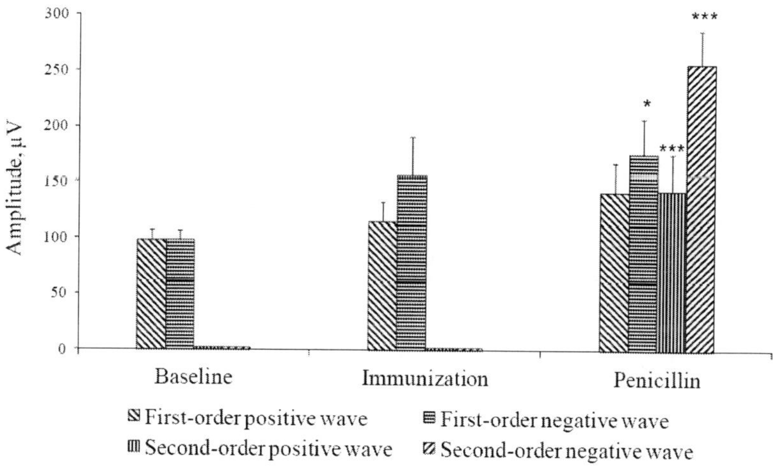

Figure 16. Changes of evoked potentials in the visual cortex of rabbits after immunization with SMAP. *p <0.05; ***p <0.001.

So, our studies show that IM administration of SMAP to rabbits suppresses epileptic seizures induced by microapplication of penicillin into the amygdala, whereas microapplication of homologous anti-SMAP antibodies into the amygdala or induction of production of anti-SMAP autoantibodies as a result of immunization of rabbits with SMAP launches epileptic seizures. Bearing in mind the antimutagenic activity of SMAP (Mekhtiev, Movsum-zadeh, 2008) and a possible underlying mechanism of conformational changes of chromatin, one may propose that the same mechanism underlies the suppressive effects of SMAP on epileptic seizures and, conversely, induction of seizure activity by intracerebral delivery of homologous or heterologous anti-SMAP antibodies. Perhaps anti-SMAP antibodies bring decondensation of chromatin structure and activation of "epileptic" genes, whereas SMAP itself leads to chromatin condensation and depression of these genes.

Conclusion

The studies presented in this chapter were conducted with a novel serotonin-modulating anticonsolidation protein (SMAP) first identified in the rat brain cortex and purified from the whole brain. The protein has an M_r of 186 kDa and is in a linear relation with the serotonin concentration (Mekhtiev, 2000; Mekhtiev et al., 2004). Studies in Wistar rats conducted in the conditioned passive avoidance and shuttle box models with electric shock reinforcement and in the instrumental alternative choice model with food reinforcement showed that intracerebral administration of SMAP prior to learning sessions impairs memory formation but has no effect if administered 24 h after learning completion (Mekhtiev, 2000; Guseinov, Mekhtiev, 2013). At the same time, SMAP administration in the reconsolidation model had no effect on memory traces.

In the morphine self-administration model, the animals achieved constant morphine intake, and the SMAP concentration was significantly upregulated in the cingulate cortex but did not change in hypothalamus. A single IM administration of anti-SMAP polyclonal antibodies to the drug-addicted animals resulted in a sharp decrease in morphine intake for more than 8 days (Mekhtiev et al., 2014).

In the experiments in *Shinshilla* rabbits, epileptic seizures were launched by microinjection of penicillin solution into the amygdala. Intramuscular administration of SMAP to rabbits after 40 min led to cessation of these seizures. Inversely, microinjection of polyclonal anti-SMAP antibodies into the amygdala of the intact rabbits initiated epileptic seizures after 2.5–3 h (Mekhtiev, Panahova, *in press*).

Thus, we can conclude that SMAP possesses anticonsolidation and anticonvulsant activities and is engaged in the regulation of drug addiction establishment.

References

Abel T, Kandel E. Positive and negative regulatory mechanisms that mediate long-term memory storage. *Brain Res. Rev.* 1998;26(2–3):360–378.

Barziali A, Kennedy TE, Sweatt JD, Kandel ER. 5-HT modulates protein synthesis and the suppression of specific proteins during long-term facilitation in *Aplysia* sensory neurons. *Neuron.* 1989;2:1577–1586.

Gasanov GG, Kuliyev EI. The neuronal activity of ventro-medial hypothalamus during pharmacological exposures that modulate motivational and emotional states. *Physiol J USSR.* 1992;78(10):29–35.

Guseinov ShB, Mekhtiev AA. Studies of the role of serotonin-modulating anticonsolidation protein in memory formation in rats in a shuttle box. *Neurosci. Behav. Physiol.* 2013;43(5):551 –556.

Hernandez EJ, Williams PA, Dudek FE. Effects of fluoxetine.

Jobe PC. Common pathogenic mechanisms between depression and epilepsy: an experimental perspective. *Epilepsy Behav.* 2003;4(Suppl. 3): S14–S24.

Kemens G, Kemens I, Michael M, Papp A, Muller U. Phase-dependent requirements for memory reconsolidation: differential roles for protein synthesis and protein kinase A activity. *J. Neurosci.* 2006;26(3):6298–6302.

Mekhtiev AA, Gasanov GG, Mehdiyev MM, Zakiyeva KF. [Effects of antibodies to serotonin-modulating protein SMP-69 on exploratory behavior in rats](Rus). *J Highest Nerv. Syst.* 1996;46(2):386–388.

Mekhtiev AA, Movsum-Zadeh SK. Antimutagenic activity of serotonergic system and underlying mechanisms in sturgeon juveniles (*Acipenser gueldenstaedti persicus*) and goldfish (*Carassius auratus*). *J. Evol. Biochem. Physiol.* 2008;44(5):476–481.

Mekhtiev AA, Rashidova AM, Muslimov IA. Suppression of morphine intake by antibodies to serotonin-modulating anticonsolidation protein in the model of self-administration of morphine. *J. Highest Nerv. Syst.* 2014;64(2):231–239.

Mekhtiev AA. Revealing in the brain of the rat the protein possessing with anticonsolidation properties. *Bull. Exp. Biol. Med.* 2000;129(8):147–150.

Nikitin VP, Solntseva S. [Protein synthesis inhibitors suppresses amnesia induced by impairment of memory reconsolidation by antagonists of receptors of glutamate and serotonin](Rus). *Russ. J. Physiol.* 2010;96(12):1223–1233.

Okuhara DY, Beck SG. 5-HT1A receptor linked to inward-rectifying potassium current in hippocampal CA3 pyramidal cells. *J. Neurophysiol.* 1994;71:2161–2167.

Parsons LH, Koob GF, Weiss F. Extracellular serotonin is decreased in nucleus accumbens during withdrawal from cocaine self-administration. *Behav. Brain Res.* 1995;73:225–228.

In: New Developments in Serotonin Research
Editor: Ming D. Li
ISBN: 978-1-63482-220-6

Chapter 7

Participation of Serotonin-Modulating Anticonsolidation Protein in Regulation of Mutagenesis and Adaptation

Arif A. Mekhtiev[*,1], *Namik J. Mustafayev*[2], *Grigoriy M. Palatnikov*[1] *and Sevda K. Movsum-zadeh*[1]
[1]A. I. Karaev Institute of Physiology, NAS of Azerbaijan, Baku, Azerbaijan
[2]Institute of Zoology, NAS of Azerbaijan, Baku, Azerbaijan

Abstract

The results of the studies presented below, were conducted with application of a novel serotonin-modulating anticonsolidation protein (SMAP) first identified in the rat brain cortex and purified from the whole brain, having M_r 186 kDa and being in linear relationship with serotonin level (Mekhtiev, 2000; Mekhtiev et al., 2004).

The adverse impact of prolonged exposure of the sturgeon juveniles in the oil-contaminated water on SMAP and mutations levels (micro-

* Corresponding author: Arif A. Mekhtiev, PhD, 2 Sharif-zadeh St., Baku, AZ1100, Azerbaijan, E-mail: arifmekht@yahoo.com.

nucleus test; Mekhtiev et al., 2010) is considered in this chapter. Besides, effects of administration of SMAP to the sturgeon juveniles exposed to the water contaminated with high levels of polyaromatic hydrocarbons and heavy metals on mutation level relatively to the control group of animals kept in the same conditions, are presented (Mekhtiev, Movsumzadeh, 2008). The effects of many-year dwelling of originally fresh-water Caspian roach, carpbream and shemaya fishes in brackish water in the Caspian Sea on SMAP level in their tissues relatively to the fishes of these species dwelling in fresh water (Mustafayev, Mekhtiev, 2013) are analyzed. Besides, the mechanism of animals' adaptation to industrial pollutants through changes in serotonergic system activity is described (Mustafayev, Mekhtiev, 2014).

Keywords: serotonin-modulating anticonsolidation protein, mutations, fish, water salinity, pollutants, cytochrom P-450, ELISA-test

Introduction

As the effects of serotonin on long-term nerve cell processes are mediated through activation of specific genes and by the synthesis of proteins being under the specific regulatory influences of serotonin (Barziali et al., 1989; Abel, Kandel, 1998) studies of the involvement of serotonin in the mechanisms of memory and in other serotonin-related processes require investigations of these proteins' participation in formation of memory traces in various conditioned behavioral models. For the first time a novel serotonin-modulating anticonsolidation protein (SMAP) was identified in the rat brain cortex with application of electrophoresis technique in the continued gradient (4-30%) of polyacrylamide gel (Mekhtiev, 2000). SMAP level was upregulated by over 84% ($p<0.01$) 24 h after administration of anti-SMP-69 protein polyclonal antibodies into the rat brain lateral ventricle. SMP-69 protein, in turn, had earlier been identified with electrophoresis technique in polyacrylamide gel after application of 10^{-4} M serotonin solution onto the occipital cortex of narcotized Wistar rats for 30 min, thereafter was purified from the whole rat brains, having M_r 69 kDa, *pI* 6.0 and being in opposite relationship with serotonin (Mekhtiev et al., 1996). SMAP was positioned in the electrophoretic lanes as a heavy-weight-molecule with low electrophoretic mobility, having M_r of 186 kDa and bearing fraction number 28. Then SMAP was purified from the whole rat brains under screening control of the ELISA-test with application of γ-globulins to electrophoretic fraction number 28. The main steps of SMAP purification include: 1) partial precipitation of water-soluble proteins in the range of 0-40% concentration of ammonium sulfate, and 2) gel-chromatography on the column of Sephadex G-150 (3 X 60 cm). For the purpose of measurement of SMAP level in the animal's tissues with the ELISA-test, anti-SMAP polyclonal γ-globulins were produced through 5-6-month immunization of rabbits with purified SMAP (300 μg per injection) always in mixture with equal amount of complete Freund adjuvant (Mekhtiev, 2000).

All experiments undertaken on animals were realized according to regulations of Directive of Council of European Society (86/609/EEC).

1. SMAP Possessing with Antimutagenic Activity

As there is enough information in the literature concerning changes of serotonergic system activity in animal tissues under impact of adverse environmental factors (Tsai et al., 1995; Handy, 2003), it gave us grounds to conduct studies of impact of chemical contamination on SMAP level in different tissues of fish. First, the level of SMAP was evaluated with the indirect ELISA-test in the livers of gobies (*Neogobius fluviatilis*), dwelling in settled way in the littoral zone of the Caspian Sea. Fish was caught in the clean zone and in the zone polluted with zinc, iron and polyaromatic hydrocarbons (PAH). It was shown that in the livers of the fish from the polluted zone of the Caspian Sea SMAP was downregulated significantly in comparison to the fish from the clean zone ($p<0.001$, Student's t-criterion). Simultaneously, mutagenic changes in the erythrocytes of the fish from the polluted zone determined by micronucleus test (Schmid, 1975), was much higher than in the fish from the clean zone ($p<0.001$, Student's t-criterion; Figure 1; Mekhtiev et al., 2010).

In the model studies realized in the sturgeon juveniles (*Acipenser gueldenstaedti persicus*) similar dynamics was revealed. The animals were culled into 3 groups: 1) intact group kept in the container with clean fresh water; 2) animals of the 1st experimental group were kept in the fresh water contaminated with oil from the sea deposit "Oil Rocks" at a concentration of 100 ppm for 5 days; 3) animals of the 2nd experimental group were kept in the fresh water contaminated with oil from the sea deposit "Oil Rocks" at a concentration of 100 ppm for 15 days. Evaluation of SMAP level (indirect solid-base ELISA-test with application of anti-SMAP IgG as first antibodies) and micronuclei amount in this fish demonstrated that while short-term exposure did not have effects on the levels of SMAP and micronuclei, prolonged exposure resulted in downregulation of SMAP (p<0.01, Student's t-criterion) and upregulation of mutations (p<0.01, Student's t-criterion; Figure 2; Mekhtiev et al., 2010). So, these data indicate to involvement of serotonergic system and, particularly, SMAP in regulation of mutagenic changes in tissues.

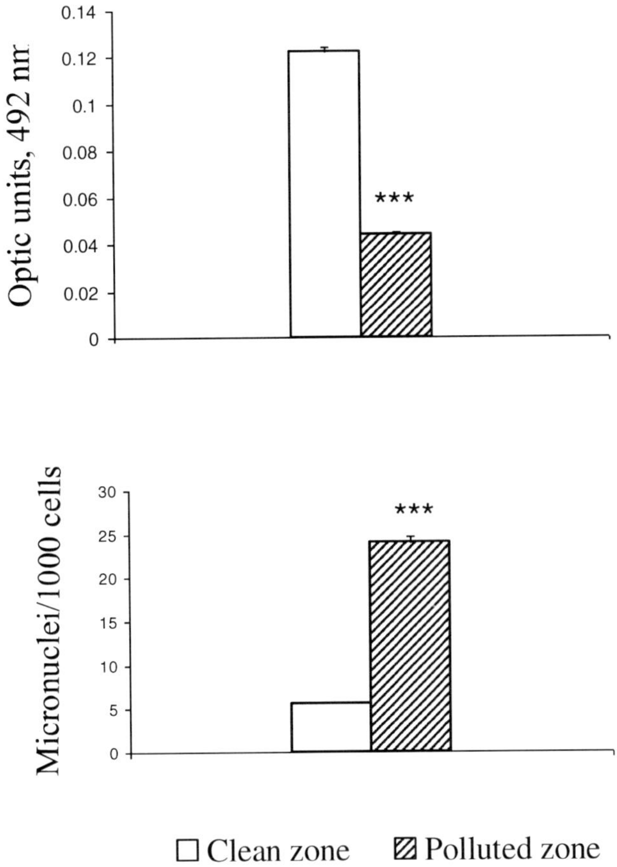

Figure 1. Levels of SMAP (ELISA-test) in the livers (A) and micronucleus amount in the erythrocytes (B) of the gobies (n=10) from the clean and polluted zones of the Caspian Sea. ***- p<0.001 on Student's t-criterion.

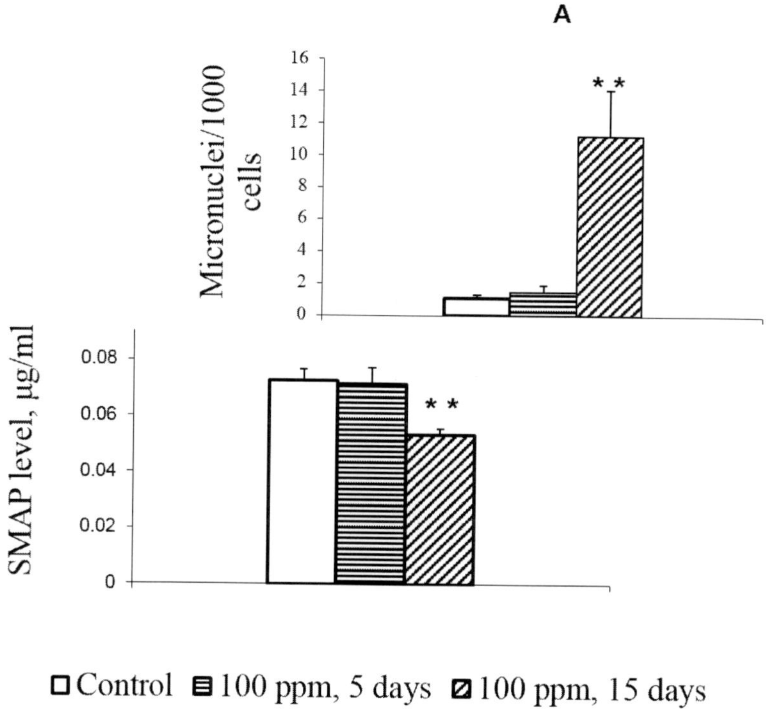

Figure 2. Impact of different time of exposure to oil on micronucleus amount in erythrocytes (A) and SMAP level in the livers (B) of the sturgeon juveniles (n=10). ** - p<0.01, on Student's t-criterion.

Next study was designed to clarify opaque role for SMAP in regulation of mutations. For this purpose animals (sturgeon juveniles) were culled into 3 groups: 1) intact group were kept in the container with clean fresh water; 2) animals were kept in the fresh water contaminated with soil sediments from Baku bay (800 ppt) containing PAH and heavy metals for 3 days and then passed to the clean water; 3) animals were kept in the contaminated water as the animals from the 2nd group, but were administered i.m. two times with SMAP (each time 1 mg). The results showed that keeping animals under impact of water contaminated with soil sediments led to five-time increase of mutagenic level (micronucleus test; p<0.001, Student's t-criterion), whereas SMAP administration brought to downregulation of mutations by over 50% (p<0.01, Student's t-criterion; Figure 3; Mekhtiev, Movsum-zadeh, 2008).

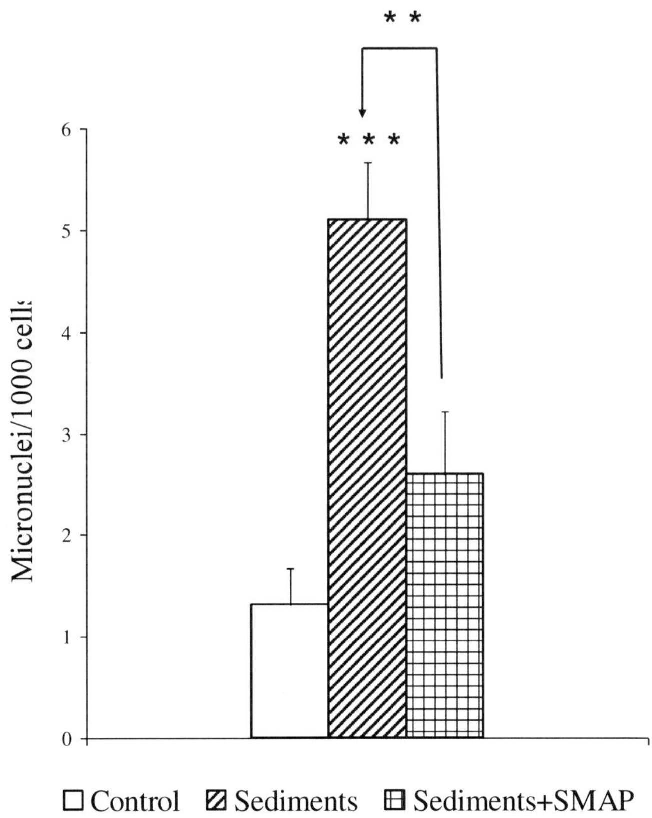

Figure 3. Impact of sediments and SMAP on the level of micronuclei in the erythocytes of sturgeon juveniles (n=10). ** - p<0,01 relatively to the "Sediments" group, *** - p<0,001 on Student's t-criterion.

To make clearer the role for SMAP in protection from mutagenic impact we studied the effects of SMAP blockade with polyclonal antibodies purified from the solution of rabbit polyclonal anti-SMAP immunoglobulins by immune affine chromatography technique in the CNBr-Sepharose (Sigma, Germany) column with the immobilized SMAP. The polyclonal anti-SMAP immunoglobulins were, in turn, produced through immunization of rabbits for 5-6 months with purified SMAP always in mixture with equal amount of complete Freund adjuvant. The studies were conducted on sturgeon juveniles of 25-40 g weight. Animals were culled into three groups: 1) intact group; 2) control group – animals were injected with rabbit non-immune γ-globulins; 3) experimental group – animals were injected with rabbit anti-SMAP. In the experimental group sturgeon juveniles were i.m. injected with anti-SMAP antibodies. Preparations were administered twice with interval of 24 h, at a concentration of 1.5 mg/ml and amount of 0.7 ml. Animals of all groups were kept in clean fresh water. The results showed that administration of anti-SMAP antibodies to the experimental animals brought to upregulation of mutagenic changes up to 56% (micronucleus test; $p<0.01$, Student's t-criterion; Figure 4; Mekhtiev et al., 2010) in comparison to the controls injected with the same amount of non-immune γ-globulins (to exclude toxic effects of heterologic γ-globulins on mutation level).

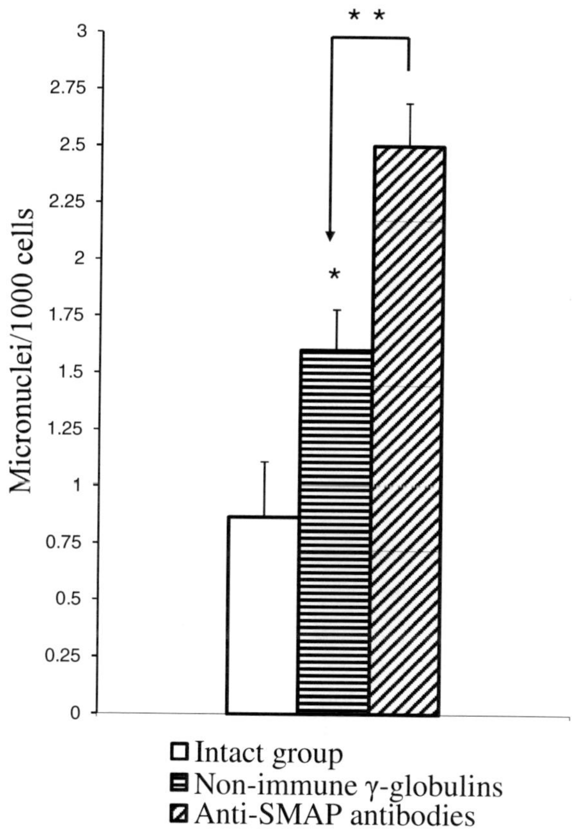

Figure 4. Impact of anti-SMAP polyclonal antibodies on micronucleus amount in the erythrocytes of the sturgeon juveniles (n=10). * - $p<0.05$; ** - $p<0.01$ on Student's t-criterion.

The results of the above presented studies indicate that serotonergic system, particularly SMAP, possesses with antimutagenic activity which realizes protection of tissues from impact of adverse environmental factors. While this system is downregulated, owing to exhaustion of its resources under impact of adverse environmental factors or other reasons, the tissues leave without antimutagenic protection and mutations level goes up. As formation of micronuclei under impact of adverse factors happens during cell division which is, probably, related to the conformational changes of chromatin structure occurring this time (Kasuba et al., 2002; Banasik et al., 2005), the observed phenomenon of mutagenesis upregulation on background of downregulation of SMAP may as well be due to such conformational switching of chromatin arranging from mainly condensed to uncondensed state being this way much more vulnerable to impact of mutagenic changes.

2. Role for SMAP in Processes of Adapatation and Disadaptation

Environmental factors of different origin induce significant changes of serotonergic system activity in animal tissues. Chronic exposure of fish in the water containing copper (Handy, 2003) and mercury (Tsai et al., 1995) contaminants brings to noticeable downregulation of serotonin in animal tissues. The studies carried out on crustacean have shown that long-term exposure of animals in the water containing contamination of heavy metals and organic substances has negative effect on serotonin turnover leading to its downregulation (Fingerman et al., 1998). Along with it, short-term exposure of rainbow trout in the water containing cadmium and lead contaminations promotes upregulation of serotonergic system in gills, liver and kidney. The goal of the present study was clarification of a pattern of changes of serotonergic system in animal tissues depending on character of the effects of different environmental factors on the animal organisms and drawing on this basis a conclusion on processes of adaptation and/or disadaptation in their tissues.

The first series of studies was carried out on vobla (*Rutilus rutilus caspicus*), bream *(Abramis brama orientalis)* and shemaya *(Chalcalburnus chalcoides)* caught in fresh (Mingechevir reservoir and brackish water (Lankaran shore and Neftchala shore of the Caspian Sea). The fish was caught in clean zones, far away from oil deposits, at the end of February-onset of March (pre-spawning period). Immediately after catch fish was sacrificed and samples of brain, liver and gills were taken from the animals and frozen.

While analyzing the water samples taken from the points of fish catches, the following data presented in the table, were obtained. The table 1 shows that over the indexes of salinity and total amount of dissolved salts, the samples taken from the shore of the Caspian Sea, overcome significantly the samples taken from reservoir.

Table 1. Values of indexes of water samples taken from the points of fish catch

Indexes	Mingechevir reservoir	Neftchala shore of the Caspian Sea	Lankaran shore of the Caspian Sea
pH	8,12	8,27	8,23
Turbidity, mg/l	15	14	16
Salinity, ‰	0,0	9	12
Total amount of dissolved salts, g/l	0,37	9	12
Special density, g/l	1000	1003	1008
Oxidation-reduction potential, mV	100	104	112

As a result of carried out comparative studies of SMAP level in the brain of vobla, bream and shemaya caught in fresh water (Mingechevir reservoir) and brackish water (Lankaran and Neftchala shores of the Caspian Sea), significant changes of SMAP levels under impact of increased salinity was revealed. Particularly, its levels in the vobla brain caught in brackish water (Lankaran and Neftchala shores of the Caspian Sea) were higher than its level in the fresh water (Mingechevir reservoir) 3.7 and 4.2 times, correspondently ($p<0.001$; Figure 5).

Similar character of changes of SMAP level was noticed in the brain of bream species: its level in the brains of fish caught in brackish water (Lankaran shore of the Caspian Sea), exceeded this index in the fish caught in fresh water by over 1.4 times ($p<0.001$; Figure 5). In the brain of shemaya upregulation of SMAP level by over 2.5 times in the animals dwelling in the sea conditions relatively to the animals dwelling in the river was also observed ($p<0.01$; Figure 5).

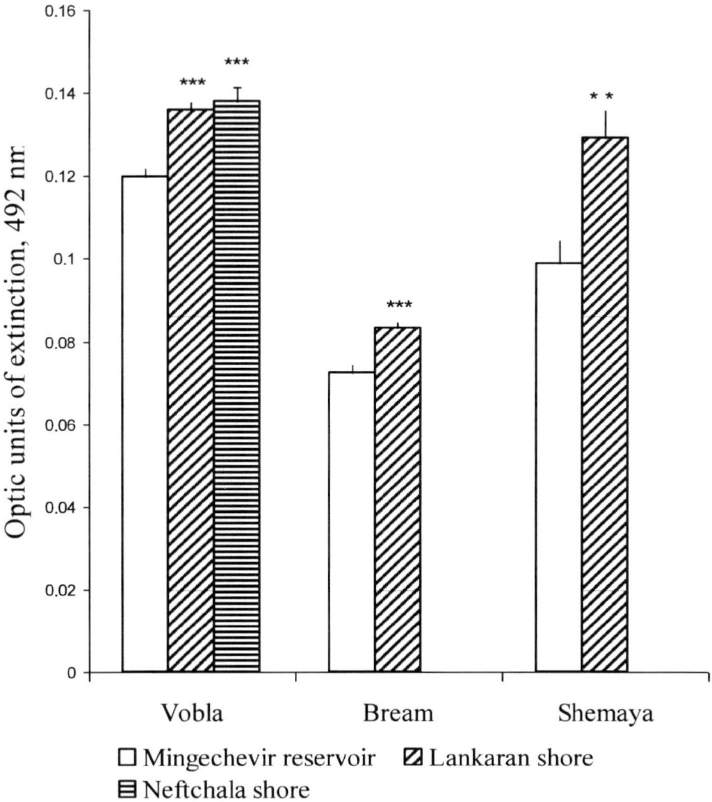

Figure 5. Level of SMAP in the brain of the fish caught in fresh and brackish waters. ** - $p<0.01$; *** - $p<0.001$ on Student's t-criterion.

Analysis of SMAP level in the livers of bream and shemaya did not reveal differences in its level in the animals caught in brackish and fresh waters. Along with it, in the bream's gills not big (4.1%; $p<0.001$), while in the shemaya's gills significant (55%; $p<0.001$; Figure 6) upregulation of SMAP level was noticed in the animals caught in the brackish water.

Besides, in laboratory conditions model experiments were conducted on the sazans *(Cyprinus carpio Linne).* In this study animals were culled into 3 groups: 1st group – animals were kept in the tanks with fresh water (n=14); 2nd group – animals were kept in the sea water taken from the clean zones of the Caspian Sea (n=15); 3rd group – animals were kept in fresh water to which sodium chloride was added up to the final concentration of 12‰ (n=16). Previously animals from 2-nd and 3rd groups were adapted to the conditions of brackish water by step-by-step elevation of salinity level of the sea water (2nd group) or fresh water (3rd group) up to 4‰ and 8‰ during 7 days each, thereafter they were placed for 21 days into the tanks with sea water or fresh water with salinity level of 12‰, correspondently.

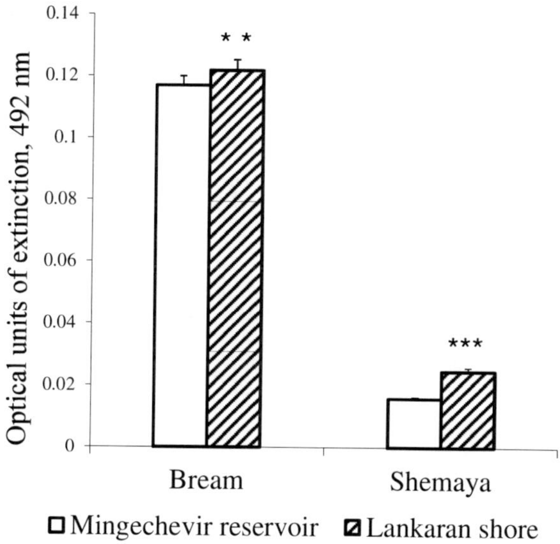

Figure 6. Level of SMAP in gills of bream and shemaya specimens caught in fresh and brackish waters (n=10). ** - p<0.01; *** - p<0.001 on Student's t-criterion.

As a result of model experiments in the sazans, significant (by over 19%) upregulation of SMAP in the brains of the animals kept in the sea (12‰) and salt (12‰) waters relatively to the animals kept in fresh water (p<0.001) was observed. At the same time in the liver and gills of sazans little upregulation of SMAP was fixed (p<0.01 and p<0.05, correspondently; Figure 7).

So, increase of salinity level of water milieu leads to significant upregulation of serotonin in the fish brains, not bringing to significant changes of its level in the other tissues.

While analyzing character of changes of SMAP level in the fish brain reflecting proportional changes of serotonin level under impact of increased salinity of environment, it should be taken into account that for the studied fish species increased water salinity is not adverse or unfamiliar factor. Owing to the fact that vobla and bream are half-migratory fishes and shemaya is migratory fish, most time of a year they dwell in the sea milieu and just in spawning time go up into the rivers.

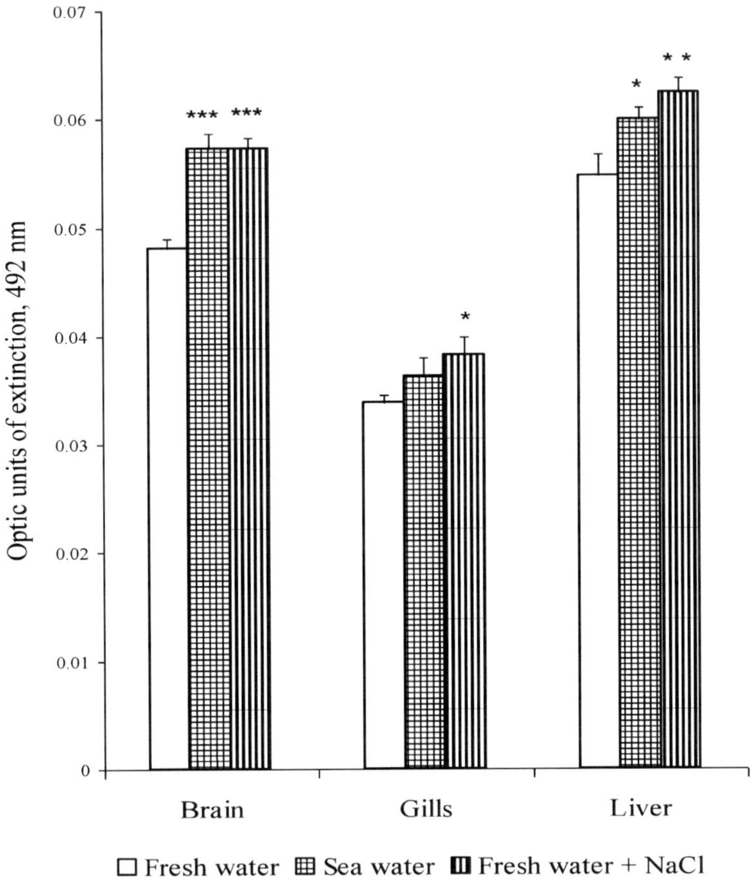

Figure 7. Changes of SMAP level in the tissues of sazan under impact of sea and "salted" fresh waters (n=10). * - $p<0.05$; ** - $p<0.01$; *** - $p<0.001$ on Student's t-criterion.

For this reason dwelling in the water of increased salinity is essential part of ecological peculiarities of animals of these species. Along with it, SMAP upregulation in the brain under increased salinity of milieu reflecting adequate upregulation of serotonergic system, indicates to launching adaptive mechanisms in the studied animals in these living conditions. The fact that changes of serotonergic system are mostly revealed in the brain, but are not observed significantly in the other studied tissues indicates to realization by the specialized brain structures the central regulation of adaptive mechanisms providing adaptation of the whole organism to dwelling in the conditions of increased salinity.

Besides the results of examination of water samples taken from the points of catches of the studied fishes, additional argument for that that just water salinity in the sea milieu was the only reason of SMAP upregulation in the brain is presented by the results of model experiments carried out in the sazans. In these experiments both the sea water and artificially "salted" fresh water induced SMAP upregulation in the animal brain to the same magnitude relatively to its values in the animals kept in fresh water, thus excluding role of any other missed factors in the said changes.

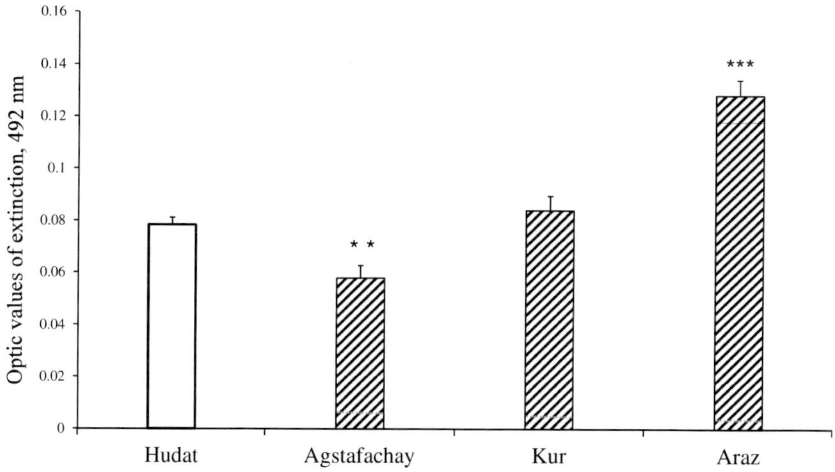

Figure 8. Level of SMAP in the gills of bystranka (n=10). ** - $p<0.01$; *** - $p<0.001$ on Student's t-criterion.

The second series of studies was conducted on eastern bystranka *(Alburnoides bipunctatus eichwaldi),* caught in the Hudat, Agstafachay, Kur and Araz rivers flowing over different districts of Azerbaijan Republic. There were 10 specimens in each group. To exclude the effects of changed levels of hormones on the studied indexes, fish was caught at December 10-12, 2012. Immediately after catch animals were sacrificed and samples of brain, liver and gills were taken from them and frozen. The contents of SMAP and cytochrome P-450 were determined in the livers, gills and brains of the studied fish by the indirect solid-phase ELISA test on the polystyrene plates. Measurement of cytochrome P-450 in fish tissues was realized in order to have a scale of harmful effects of adverse environmental factors.

In the second series of studies the following regularities were found. SMAP levels in the gills and liver of bystranka caught in the rivers of Hudat, Agstafachay, Kur and Araz flowing over territory of Azerbaijan, had similar pattern. Particularly, SMAP level in the gills and liver of bystranka caught in the Agstafachay river, was significantly lower (by over 26% and 45%, correspondently; $p<0.01$; Figure 8 and 9) than in the animals from the Hudat river, being, according to the data of Ministry of Ecology of Azerbaijan Republic (Mamedova, 2012), a zone free from industrial and domestic pollutants. On the other hand, SMAP level in the gills and liver of the fish caught in the Kur river, was significantly higher (by over 64% and 43%, correspondently) than in the fish from the Hudat river ($p<0.001$; Figure 8 and 9). SMAP was as well upregulated by over 29% ($p<0.01$) in the liver of the fish from the Kur river.

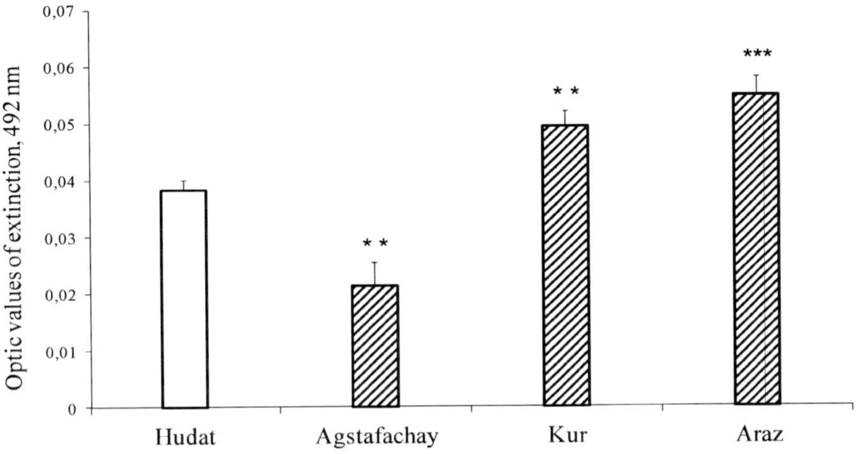

Figure 9. Level of SMAP in the liver of bystranka (n=13). ** - $p<0.01$; *** - $p<0.00$ on Student's t-criterion.

Measurement of the level of biomarker cytochrome P-450 in the gills and liver of the fish caught in the said rivers, revealed its sharp downregulation in the liver of animals from the Agstafachay river (by over 69%; $p<0.001$), Kur river (by over 39%; $p<0.001$) and Araz river (by over 28%; $p<0.01$; Figure 10) relatively to its level in the liver of the animals from the Hudat river. In the gills of the fish from the Agstafachay river significant downregulation (by over 29%; $p<0.001$) of cytochrome P-450 and little (by over 16%) its downregulation in the gills of the fish from the Kur river relatively to the said index in the fish from the unpolluted river (Figure 11) was revealed.

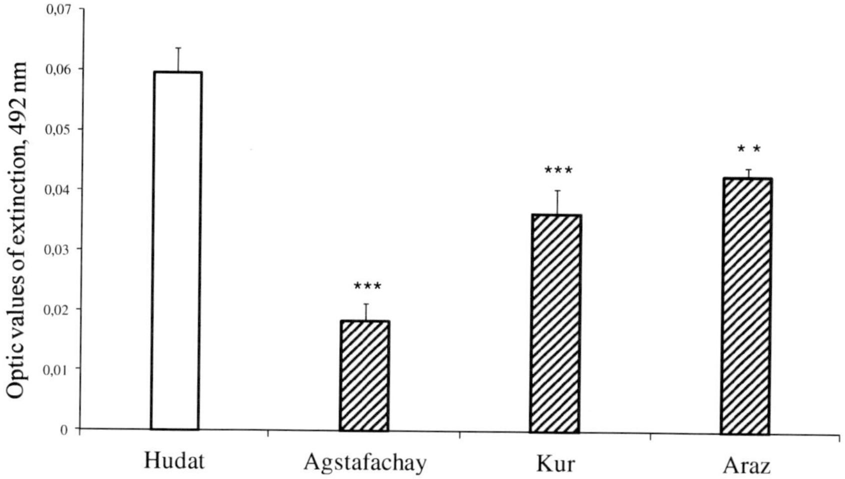

Figure 10. Level of Cytochrome P-450 in the liver of bystranka (n=13). ** - p<0.01; *** - p<0.001 on Student's t-criterion.

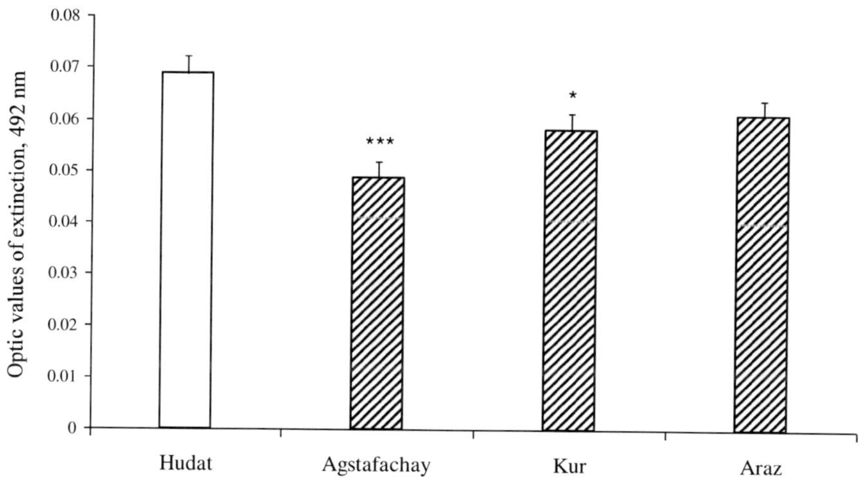

Figure 11. Level of Cytochrome P-450 in the gills of bystranka (n=13). * - p<0.05; *** - p<0.001 on Student's t-criterion.

Different results were obtained while measuring SMAP level in the brain of bystranka. In the fish caught in the Kur and Araz rivers SMAP upregulated (by over 17% and 23%, correspondently; p<0.001), whereas its level in the fish from the Agstafachay river did not differ from the values in the fish from the Hudat river (Figure 12). While measuring level of cytochrome P-450 in the brain, its slight (by over 14%) downregulation in the fish from the Agstafachay river was observed ($p<0.05$).

So, while determining levels of SMAP and cytochrome P-450 in the liver and gills of bystranka, significant changes of their values were revealed in the fish caught in the polluted rivers of Agstafachay, Kur and Araz relatively to the values of the fish from the unpolluted Hudat river, while their levels in the fish brain were not so indicative for pollutants' impact.

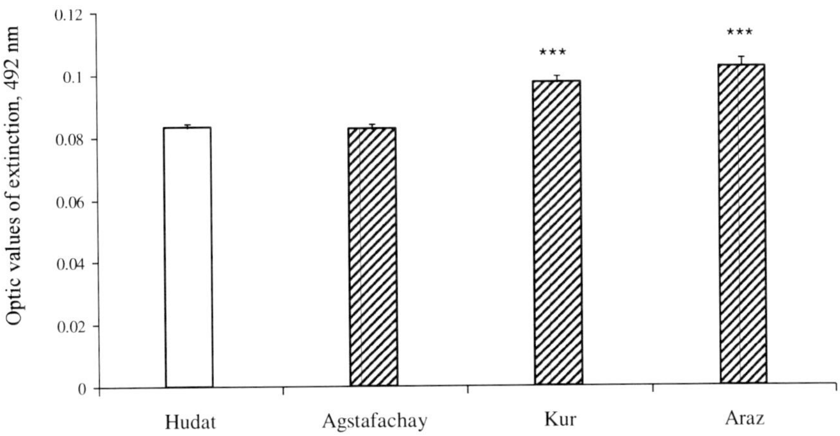

Figure 12. Level of SMAP in the brain of bystranka (n=13). *** - $p<0.001$ on Student's t-criterion.

Discussion

Consideration of the level of biomarker cytochrome P-450 in the tissues of the specimens of bystranka caught in the Hudat, Agstafachay, Kura and Araz rivers gives grounds to us to reveal degree of adverse impact of pollutants of these rivers on functioning of the studied tissues. Sharp downregulation of cytochrome P-450 in the liver and gills of the fish caught in the Agstafachay river relatively to its level in the fish from the Hudat river, unpolluted with industrial and domestic wastes, indicates to presence of pollutants, having harmful effects on the fish organism, in the water of this river. The revealed changes in the level of cytochrome P-450 caught in the Agstafachay river, coincide with the analogous changes of SMAP level in the liver and gills of these fish. According to the data of Ministry of Ecology of Azerbaijan Republic, phenol level in the water of this river 4 times exceeds the maximal adopted level (Mamedova, 2012). Our earlier conducted laboratory studies showed that long-term exposure of sturgeon juveniles in the water polluted with oil and industrial wastes, as well as dwelling of gobies in the littoral zone of the Caspian Sea polluted with industrial wastes (Mekhtiev et al., 2010) leads to significant downregulation of SMAP in the liver samples accompanied with sharp increase of mutagenic changes in the tissues of these animals. The results of the studies on inhibition of mutagenic changes in tissues of sturgeon juveniles kept in the water contaminated with heavy metals and polyaromatic hydrocarbons (Mekhtiev, Movsum-zadeh, 2008), indicate that downregulation of SMAP under long-term exposure of animals in the water contaminated with pollutants does not have accompanying or accidental character, but reflects exhaustion of resources of serotonergic system of the organism realizing antimutagenic activity. So, sharp downregulation of cytochrome P-450 and SMAP in the liver and gills of the fish caught in the Agstafachay river, makes an evidence of profound negative impact of the pollutants being in the water, on their organisms.

While analyzing changes of levels of cytochrome P-450 and SMAP in the liver and gills of the fish caught in the Kura and Araz rivers, one can see opposite directional changes of these indexes. Particularly, in the fish caught in the Kur and Araz rivers significant downregulation of cytochrome P-450 in the liver was observed, while in the fish from the Kur river its downregulation in the gills was noticed. At the same time in the livers and gills of the fish caught in the Araz river as well as in the livers of fish from the Kur river significant upregulation of SMAP was revealed.

Taking into account the data on SMAP upregulation in the tissues of vobla, bream and shemaya dwelling for a long time in brackish water relatively to its level in the tissues of the fish of these species from fresh-water milieu, indicating to adaptive character of these changes, one may come to a conclusion that above noticed SMAP upregulation in the liver and gills in the fish from Araz river and its upregulation in the liver of the fish from the Kur river has compensatory character and shows adaptation of bystranka specimens to a range of the pollutants presenting in these rivers.

Analysis of changes of the SMAP level in the brain of the fish caught in the Kur and Araz rivers revealing its sharp upregulation relatively to the values of the fish from the Hudat river, from our standpoint, confirms compensatory character of changes of this protein level in the liver and gills of the fish from the Araz river and in the liver of the fish from the Kur river. Along with it, SMAP level in the brain of the fish caught in the Agstafachay river, in contrast to the character of their changes in the liver and gills of these fish, does not differ from its level in the fish from the unpolluted river. On the other hand, with exclusion of the fish caught in the Agstafachay river (slight downregulation) the level of cytochrome P-450 in the brains of the fish from the all studied rivers does not change from its level in the Hudat river. These data, apparently, show that brain does not respond strongly to impact of adverse environmental factors (perhaps, due to existence of blood-brain barrier), and revealed changes in levels of single proteins under these factors' impact are rather related to realization by brain regulatory functions regarding to other organs and tissues.

So, the data got in the present study, give grounds to make a conclusion that upregulation of SMAP (serotonergic system, correspondently) under impact of different environmental factors (including harmful ones) indicate to adaptation of the organism's systems to these factors, whereas its downregulation (i.e., exhaustion of serotonergic system resources) might serve as an index of disadaptation of the organism to the effects of adverse environmental factors.

Conclusion

The results of the studies presented above, were conducted with application of a novel serotonin-modulating anticonsolidation protein (SMAP) first identified in the rat brain cortex and purified from the whole brain, having M_r 186 kDa and being in linear relationship with serotonin level (Mekhtiev, 2000; Mekhtiev et al., 2004).

Prolonged exposure of sturgeon juveniles in the oil-contaminated water in containers brings to downregulation of SMAP (exhaustion of serotonergic system resources) with significant upregulation of mutations (micronucleus test; Mekhtiev et al., 2010). Administration of SMAP to sturgeon juveniles (i.m.) leads to two times downregulation of mutation level in animals exposed to polyaromatic hydrocarbons and heavy metals relatively to the animal group kept in the same adverse conditions (Mekhtiev, Movsum-zadeh, 2008).

Many-year dwelling of originally fresh-water Caspian roach, carpbream and shemaya fishes in brackish water in the Caspian Sea results in adaptive upregulation of SMAP level in their tissues relatively to its level in the fishes of the same species dwelling in fresh water (Mustafayev, Mekhtiev, 2013). The same kind of water animals' adaptation is shown in regard to industrial pollutants (Mustafayev, Mekhtiev, 2014).

So the results show that SMAP possesses with antimutagenic and adaptogenic activities.

References

Abel T., Kandel E. Positive and negative regulatory mechanisms that mediate long-term memory storage. *Brain Res. Rev.* 1998. 26 (2–3): 360–378.

Banasik A., Lankoff A., Piskulak A., Adamowska K., Lisowska H., Wojcik A. Aluminum-induced micronuclei and apoptosis in human peripheral-blood lymphocytes treated during different phases of the cell cycle // *Environ. Toxicol.* 2005. V. 20, № 4, P. 402-406.

Barziali A., Kennedy T.E., Sweatt J.D., Kandel E.R. 5-HTmodulates protein synthesis and the suppression of specific proteins during long-term facilitation in Aplysia sensory neurons. *Neuron.* 1989. 2: 1577–1586.

Bickham J.W., Rowe G.T., Palatnikov G.M. , Mekhtiev A.A., Mekhtiev M., Kasimov R.Y., Hauschultz D.W., Wickliffe J.K., Rogers W.J. Acute and genotoxic effects of Baku

Harbor sediment on Russian sturgeon, *Acipenser guildensteidti.* Bull. *Environ. Contam. Toxicol.* 1998, V. 61, pp. 512–518.

Fingerman M., Jackson N. C. and Nagabhushanam R. Hormonally-regulated functions in crustaceans as biomarkers of environmental pollution. Comparative Biochemistry and Physiology. *Part C: Pharmacology, Toxicology and Endocrynology*, 1998, V. 120, № 3, pp. 343-350.

Handy R.D. Chronic effects of copper exposure versus endocrine toxicity: two sides of the same toxicological process? *Comparative Biochemistry and Physiology – Part A: Molecular & Integrated Physiology,* 2003, V. 135, № 1, pp. 25-38.

Kasuba V., Rozgaj R.. Micronucleus distribution in human peripheral blood lymphocytes treated in vitro with cadmium chloride in G0 and S phase of the cell cycle// *Chemosphere.* 2002. V. 49, № 1, P. 91-95.

Mamedova G.M. Report of the department of geochemical regime and monitoring of pollution of natural waters of Azerbaijan Republic on February, 2012 y. 2012.

Mekhtiev A.A. Revealing in the brain of the rat the protein possessing with anticonsolidation properties. *Bulletin Exper. Biol. Med.* 2000. 129(8): 147-150.

Mekhtiev A.A., Movsum-zadeh S.K. Antimutagenic activity of serotonergic system and underlying mechanisms in sturgeon juveniles (*Acipenser gueldenstaedti persicus*) and goldfish (*Carassius auratus*). *Journal Evol. Biochem. Physiol.* 2008. 44(5): 476-481.

Mekhtiev A.A., Palatnikov G.M., Movsum-zadeh S.K., Kasimov R.Yu. Upregulation of mutations in the tissues of gobies and sturgeon juveniles under antibodies-mediated blockade of serotonin-modulating anticonsolidation protein. *Journal Evol. Biochem. Physiol.* 2010. 46(5): 375-379.

Tsai C.L., Jang T.H., Wang L.H. Effects of mercury on serotonin concentration in the brain of tilapia, Oreochromis mossambicus. *Neurosci. Lett.*, 1995, 194(3):.208-211.

Index

B

C

D

E

F

I

J

K

L

M

N

R

S

T

U

V

The Law of Securitisations

The book *The Law of Securitisations: From Crises to Techno-sustainability* provides a full and detailed account of the EU legislation in the area of structured finance with the new legal rules dissected and discussed in their full extent. Securitisation transactions have been identified in the literature among the main reasons for the 2007–2008 financial crisis, alongside derivative contracts. More than a decade later, the EU legislature passed in 2017 a legal framework comprehensively disciplining the area of securitisations in the EU. On such a background the main purpose of the book is to discuss and analyse, in a holistic way, both the rationale behind the securitisations as financial transactions and their main players (e.g. originators, SPVs and credit rating agencies) and their "ESG" (Environmental, Social and Governance) challenges, particularly the recent regulation passed in the EU during the 2020–2021 global pandemic. The goal of this legal analysis is to identify and clarify the entire legal process of securitisations, as a result of the new EU legislation, as well as duties, responsibilities and practices incumbent on the main players. Furthermore, the monograph is also concerned with the new challenges facing financial markets and their regulation: the new concept of sustainability and the development of technology. In this scenario, there is a blend of financial issues, new environmental challenges and, ultimately, the role human beings are expected to play, also from a social justice perspective.

Adopting not just doctrinal methodology but also comparative (from a private law perspective) and interdisciplinary (regulatory and law and economics), the authors also include a discussion of the main literature which has blossomed over the last two decades on structured finance transactions, particularly the literature that unveiled, a decade ago, the concept of shadow banking. This book will be one of the first to focus on the new EU Securitisation Regulation and will be of interest to academics, students and practitioners of financial law.

Pierre de Gioia Carabellese (JD cum laude, PhD, LLM PGCAP, CMr), Professor (Chair) in England and Wales (Huddersfield, School of Law, 2017), has become Professor (full) of Business Law and Regulation in Australia (ECU, Perth, 2020) and Professor (full) of Banking and Financial Law in China (Beijing Institute of Technology, School of Commercial Law, Zhuhai Campus, Hong Kong Area, 2021). Professor de Gioia Carabellese is also a Notary Public in Edinburgh and a qualified lawyer (Solicitor & Avvocato) in both the United Kingdom and Italy.

Camilla Della Giustina is a PhD candidate in Law at Campania University, Luigi Vanvitelli, Italy, and she holds a Juris Doctor cum laude at the University of Padova Law School. She has already contributed 50 research papers, 2 monographs and 1 book.

Routledge Research in Finance and Banking Law

The European Sovereign Debt Crisis
Breaking the Vicious Circle between Sovereigns and Banks
Phoebus L. Athanassiou and Angelos T. Vouldis

Global Finance in the 21st Century
Stability and Sustainability in a Fragmenting World
Steve Kourabas

Banking Regulation in Africa
The Case of Nigeria and Other Emerging Economies
Folashade Adeyemo

Crypto-Finance, Law and Regulation
Governing an Emerging Ecosystem
Joseph Lee

Regulation and Governance of Mutual Funds
United Kingdom and United States of America Perspectives on Investor Protection
Mohammed Khair Alshaleel

Regulation of Debt Collection in Europe
Understanding Informal Debt Collection Practices
Cătălin-Gabriel Stănescu

The Regulation of Megabanks
Global Systemically Important Banks and their Legal Frameworks
Katarzyna M. Parchimowicz

Regulating Open Banking
Comparative Analysis of the EU, the UK and Taiwan
Chang-Hsien Tsai and Kuan-Jung Peng

For more information about this series, please visit: www.routledge.com/Routledge-Research-in-Finance-and-Banking-Law/book-series/FINANCIALLAW

The Law of Securitisations
From Crises to Techno-sustainability

Pierre de Gioia Carabellese and Camilla Della Giustina

Routledge
Taylor & Francis Group
LONDON AND NEW YORK

First published 2023
by Routledge
4 Park Square, Milton Park, Abingdon, Oxon OX14 4RN

and by Routledge
605 Third Avenue, New York, NY 10158

Routledge is an imprint of the Taylor & Francis Group, an informa business

British Library Cataloguing-in-Publication Data
A catalogue record for this book is available from the British Library

ISBN: 978-0-367-26248-8 (hbk)
ISBN: 978-1-032-45081-0 (pbk)
ISBN: 978-0-429-29343-6 (ebk)

DOI: 10.4324/9780429293436

Typeset in Times New Roman
by Deanta Global Publishing Services, Chennai, India

This book is dedicated by Professor Pierre de Gioia Carabellese to his sister, Nunzia de Gioia Carabellese, for her intelligence, grace and beauty.
This book is dedicated by Dottoressa Camilla Della Giustina to her mother.